Nordic Wittgenstein Studies

Volume 3

Series Editor
Niklas Forsberg (Uppsala University)

Editorial Board
Sorin Bangu (University of Bergen)
Martin Gustafsson (Åbo Akademi University)
Kjell S. Johannessen (University of Bergen)
Oskari Kuusela (University of East Anglia)
Yrsa Neuman (Åbo Akademi University)
Bernt Österman (University of Helsinki)
Alois Pichler (University of Bergen)
Simo Säätelä (University of Bergen)
Anne-Marie Søndergaard Christensen (University of South Denmark, Odense)
Sören Stenlund (University of Uppsala)
Thomas Wallgren (University of Helsinki)
Cato Wittusen (University of Stavanger)

Advisory Board
Maija Aalto-Heinilä (University of Eastern Finland)
Hanne Appelqvist (University of Turku)
Avner Baz (Tufts University)
Anat Biletzki (Tel Aviv University and Quinnipiac University)
Steen Brock (Aarhus University)
Kevin Cahill (University of Bergen)
David Cockburn (University of Wales)
James Conant (University of Chicago)
Cora Diamond (University of Virginia)
Alberto Emiliani (University of Helsinki)
Juliet Floyd (Boston University)
Gottfried Gabriel (Friedrich-Schiller-Universität Jena)
Dinda L. Gorlée (The Hague, The Netherlands)
Herbert Hrachovec (University of Vienna)
Allan Janik (University of Innsbruck)
James Klagge (Virginia Tech)
Michael Kremer (University of Chicago)
Camilla Kronqvist (Åbo Akademi University)
D. K. Levy (University of Edinburgh)
Denis McManus (University of Southampton)
Felix Mühlhölzer (Georg-August Universität Göttingen)
Jean-Philippe Narboux (Université Bordeaux Montaigne)
Joachim Schulte (Universität Zürich)
Daniele Moyal-Sharrock (University of Hertfordshire)
Stephen Mulhall (New College, University of Oxford)
Antonia Soulez (Université de Paris 8)
David G. Stern (University of Iowa)
Nuno Venturinha (Nova University of Lisbon)
David E. Wellbery (University of Chicago)
Edward Witherspoon (Colgate University, New York)

The series publishes high-quality studies of Ludwig Wittgenstein's work and philosophy. It is affiliated with The Nordic Wittgenstein Society, The Wittgenstein Archives at the University of Bergen and The von Wright and Wittgenstein Archives at the University of Helsinki. The series welcomes any first rank study of Wittgenstein's philosophy, biography or work, and contributions in the subject areas of philosophy and other human and social studies (including philology, linguistics, cognitive science and others) that draw upon Wittgenstein's work. It also invites studies that demonstrate the philosophical relevance of Wittgenstein's Nachlass as well as purely philological or literary studies of the Nachlass. Each submission to the series, if found eligible by the series editor, is peer reviewed by the editorial board and independent experts. The series accepts submissions in English of approximately 80 000 – 125 000 words. For further information (about how to submit a proposal, formatting etc.), please contact: niklas.forsberg@filosofi.uu.se.

More information about this series at http://www.springer.com/series/13863

Gisela Bengtsson • Simo Säätelä • Alois Pichler
Editors

New Essays on Frege

Between Science and Literature

Editors
Gisela Bengtsson
Department of Philosophy
Uppsala University
Uppsala, Sweden

Simo Säätelä
Department of Philosophy
The University of Bergen
Bergen, Norway

Alois Pichler
Department of Philosophy
The University of Bergen
Bergen, Norway

ISSN 2520-1514 ISSN 2520-1522 (electronic)
Nordic Wittgenstein Studies
ISBN 978-3-319-89047-0 ISBN 978-3-319-71186-7 (eBook)
https://doi.org/10.1007/978-3-319-71186-7

© Springer International Publishing AG 2018
Softcover re-print of the Hardcover 1st edition 2018
This work is subject to copyright. All rights are reserved by the Publisher, whether the whole or part of the material is concerned, specifically the rights of translation, reprinting, reuse of illustrations, recitation, broadcasting, reproduction on microfilms or in any other physical way, and transmission or information storage and retrieval, electronic adaptation, computer software, or by similar or dissimilar methodology now known or hereafter developed.
The use of general descriptive names, registered names, trademarks, service marks, etc. in this publication does not imply, even in the absence of a specific statement, that such names are exempt from the relevant protective laws and regulations and therefore free for general use.
The publisher, the authors and the editors are safe to assume that the advice and information in this book are believed to be true and accurate at the date of publication. Neither the publisher nor the authors or the editors give a warranty, express or implied, with respect to the material contained herein or for any errors or omissions that may have been made. The publisher remains neutral with regard to jurisdictional claims in published maps and institutional affiliations.

Cover illustration: The cover makes use of Wittgenstein Nachlass MS 115, page 118. The Master and Fellows of Trinity College Cambridge and the University of Bergen have kindly permitted the use of this picture.

Printed on acid-free paper

This Springer imprint is published by Springer Nature
The registered company is Springer International Publishing AG
The registered company address is: Gewerbestrasse 11, 6330 Cham, Switzerland

Preface

The aim of this collection of previously unpublished essays is to address the relation between philosophy, *Wissenschaft* (in the sense of strict or natural science), and *Dichtung* (i.e., literature, fiction, or poetry) in the work of Gottlob Frege. What does Frege say about *Dichtung* in relation to logic and philosophy? What does this tell us about Frege's conception of philosophy and its relation to the sciences? Does Frege himself use elements of *Dichtung* in his writings? To which purpose? An important point of comparison here is Wittgenstein, who, in addition to listing Frege as one of his main influences, famously claimed that "Philosophie dürfte man eigentlich nur dichten." The complex relation between Frege and Wittgenstein is addressed in several of the contributions, and two of the authors (Janik and Venturinha) have a particular focus on Wittgenstein. Our hope is that this anthology will provide stimulating new perspectives on these questions and nuance the received picture of Frege as a strict logicist and propagator of a scientific view of philosophy.

This anthology has its origin in the international workshop *Frege zwischen Dichtung und Wissenschaft*, held at the Wittgenstein Archives and the Department of Philosophy at the University of Bergen in December 2014, organized by the editors. We want to thank all the participants of that workshop for stimulating presentations and discussions. In addition to papers originally presented at the workshop, the anthology comprises three papers—by Korhonen, Polimenov, and Venturinha—invited by the editors.

We would like to thank the Department of Philosophy at the University of Bergen and the Austrian Embassy in Oslo for financial support of the workshop back in 2014.

Uppsala, Sweden
Bergen, Norway

Gisela Bengtsson
Alois Pichler
Simo Säätelä

July 2017

Contents

1 Introduction: Zwischen Dichtung und Wissenschaft............. 1
  Gisela Bengtsson

2 Science and Fiction: A Fregean Approach 9
  Gottfried Gabriel

3 Frege's Unquestioned Starting Point: Logic as Science 23
  Jan Harald Alnes

4 Frege, the Normativity of Logic, and the Kantian Tradition 47
  Anssi Korhonen

5 Frege's Critique of Formalism................................ 75
  Sören Stenlund

6 Why Is Frege's Judgment Stroke Superfluous?.................. 87
  Martin Gustafsson

7 Frege on Dichtung and Elucidation........................... 101
  Gisela Bengtsson

8 Semantic and Pragmatic Aspects of Frege's Approach
  to Fictional Discourse 119
  Todor Polimenov

9 The Dichtung of Analytic Philosophy: Wittgenstein's Legacy
  from Frege and Its Consequences 143
  Allan Janik

10 Agrammaticality 159
   Nuno Venturinha

Author Index 177

Subject Index.. 181

Contributors

Jan Harald Alnes UiT The Arctic University of Norway, Tromsø, Norway

Gisela Bengtsson Uppsala University, Uppsala, Sweden

Gottfried Gabriel The Friedrich Schiller University Jena, Jena, Germany

Martin Gustafsson Åbo Akademi University, Turku, Finland

Allan Janik University of Vienna, Vienna, Austria
University of Innsbruck, Innsbruck, Austria

Anssi Korhonen University of Helsinki, Helsinki, Finland

Alois Pichler University of Bergen, Bergen, Norway

Todor Polimenov Sofia University, Sofia, Bulgaria

Sören Stenlund Uppsala University, Uppsala, Sweden

Simo Säätelä University of Bergen, Bergen, Norway

Nuno Venturinha Universidade Nova de Lisboa, Lisbon, Portugal

Chapter 1
Introduction: Zwischen Dichtung und Wissenschaft

Gisela Bengtsson

"Simple, forceful, strict" are the words Georg Henrik von Wright uses to describe Gottlob Frege's style of writing (von Wright 1993, 60). He adds that it often contains an element of ice-cold irony, and this description seems to capture well the style that had such a great impact on Ludwig Wittgenstein's sentences (cf. Wittgenstein 1981, § 712). In a later essay, von Wright (1994) borrows a distinction between two different human intellectual approaches from Friedrich Waismann (1940), and gives it a central role in an outline of the origin and development of analytic philosophy. The distinction is between a scientific approach that has the search for knowledge and true propositions as a primary guideline, and a philosophical approach that views clarity as the ultimate goal. Those guided by a philosophical approach seek to make clear what propositions mean. Characteristic of this approach is the conception that philosophy is distinct from science, as it neither is directed at the construction of theories, nor guided by a search for knowledge in the form of true propositions. The scientific approach, on the other hand, is characterized by a unified view of science according to which philosophy forms part of it. Interaction and conflict between these two approaches characterize the development of analytic philosophy, according to the picture von Wright presents. Without hesitation, he lets Bertrand Russell represent the first approach and G. E. Moore the second. Frege is spoken of much more cautiously. It is as if von Wright does not quite know what to say or where to place Frege with regard to the distinction between the two different intellectual approaches.

It may seem as if the answer to where he belongs is ready at hand: No matter how we conceive of Frege's project, we will find logic, and the development of a language that is in agreement with it, at its core. Truth points the way for logic, Frege writes, and truth is what we seek in scientific investigations. He speaks of the laws of logic as the laws of truth, and it is evident that truth is at the center of his work

G. Bengtsson (✉)
Uppsala University, Uppsala, Sweden
e-mail: gisela.bengtsson@filosofi.uu.se

(Frege 1918, 58/1997, 325). We should be able to say with determination that Frege's approach is scientific, with regard to the distinction discussed earlier. In his outline, Von Wright also relies on a distinction made by Max Black in "Relations between Logical Positivism and the Cambridge School of Analysis" (1939): Black suggests that it is possible to divide philosophers into those who ask "Is it true?", and those who ask "What does it mean?" If we take Frege's approach to be (purely) scientific, he should be among those who try to answer the first question. However, knowing how urgent Frege finds the latter question it is difficult to settle with this conclusion. Von Wright is somewhat vague on this point. Turning to the question of who the originator of analytic philosophy was, he writes that Frege was the first to use the logical-analytic method on philosophical questions, but adds that it was through Russell's efforts to emphasize the importance of Frege's work that the latter came to be so widely known. He finally suggests that Frege and Russell should both be seen as the originators of analytic philosophy (von Wright 1994, 6).

It cannot be doubted that Frege's work has had major impact on contemporary philosophy. His approach to logic, mathematics and philosophy of language was bold, and his groundbreaking work in these fields form a starting point for analytic philosophy. It has also been argued that Frege is the philosopher to turn to when seeking to understand Wittgenstein's approach to philosophical difficulties and his struggle to find a form of presentation for his investigations. This legacy is acknowledged by Wittgenstein in different ways throughout his authorship. Even though Frege's great significance for contemporary philosophy is not disputed, the question of how we are to understand the character and aim of his project is difficult to answer. A point of disagreement among interpreters is Frege's conception of logic; he describes logic as the most general science and presents it as a universal framework for thinking. Since he appears to give logic an all-embracing status, it is common to conceive of him as a scientist, and further, as a proponent of a scientistic view of philosophy. The understanding of logic that Frege works with implies (according to many interpreters) an exclusion of a stance external to logic, from which to study and theorize about it. It is therefore difficult to decide how we are to conceive of the status of some of Frege's most famous articles, such as "Über Sinn und Bedeutung": does Frege present a semantic theory here, or is the presentation of a theory about the relation between language, thinking, and the world incompatible with his conception of logic? It has been suggested that Frege's discursive and philosophical writings on linguistic meaning best can be seen as consisting (to a large extent) of prose that serves didactic purposes. According to this view, Frege uses a form of prose that cannot form part of the system of science to shed light on crucial features of the purely logical language that he introduces. In this propaedeutic prose Frege offers hints and figurative forms of language, as definitions of that which is logically primitive cannot be provided. It is interesting to compare his mode of writing in the discursive prose to the manner in which he draws a distinction between the strict sciences and the humanities—*die Geisteswissenschaften*. The latter are closer to *Dichtung* than the former, in Frege's view. To clarify, he points to the different degrees of difficulty involved in translating work within different disciplines, from one language to another: on a scale of such difficulties, we

1 Introduction: Zwischen Dichtung und Wissenschaft

will find scientific expositions in the strict sciences at one end and poetry at the other. Poetry and literature—*Dichtung*—is almost impossible to translate perfectly, according to Frege, since elements that are not covered by assertoric force are abundant in this form of discourse, and the nationality of the author will be discernable. The approach of *Dichtung* is by way of intimation, and what one wishes to approach cannot always be grasped in thought, he writes. The scale of translatability correlates with different degrees in which something is scientific, as Frege presents it— clearly, *Dichtung* is unscientific, and the humanities are less scientific than the strict sciences that are "drier in proportion to being more exact" (Frege 1918, 63/1997, 330.) The way in which he draws this distinction fits well with the description of his discursive prose as 'propaedeutic'.

To shed some further light on how Frege's discursive writings may be understood, we can look at a comparison Joan Weiner (2001) suggests. If we take it, she writes, that Frege when offering hints, metaphors and other forms of elucidations, uses sentences that (from a logical point of view) are nonsensical, we may ask whether the value of this nonsense is exhausted by its use in introducing the logical notation. She goes on to say that it is not very difficult to introduce a logical notation to undergraduates, and points out that most undergraduates are taught a version of Frege's first order logical notation. What role can Frege's eludications have, Weiner asks, after someone's has caught on to the logical language and is able to use it? Is there not something that goes beyond, or comes in addition to, acquiring the ability to use the logical language with understanding? To answer this question, Weiner directs us to the penultimate remark of Wittgenstein's *Tractatus* that tells us that a reader who has used Wittgenstein's propositions as elucidations will come to see the word rightly. Her suggestion is that Frege, by using elucidations in the discursive writings, aims at transforming our way of looking at language and arithmetic. Such a new way of looking cannot be expressed in statements, nor can a transformation of our conception of language and mathematics as a whole be effected by way of arguments and statements of fact. That Frege's discursive writings, for instance in *The Foundations of Arithmetic*, are directed at a goal of this kind sheds light on their value, and this is what Weiner seeks to demonstrate (Weiner 2001, 53). Weiner's suggestion offers a way of looking at Frege's discursive writings that would support the view that his approach there is philosophical rather than scientific—with regard to the distinction spoken of earlier.[1]

In a paper from 2013, Jan Harald Alnes points to the fact that disagreements between different interpretative traditions within Frege research often have remained unresolved over time. The lack of agreement may seem surprising, Alnes writes, in view of the general agreement that Frege's writings represent an ideal when it comes to clarity (Alnes 2013, 257). A possible source of disagreement, he suggests, is that interpreters tend to focus on separate areas of Frege's comprehensive project and that the focus of debates often are isolated questions. Alnes writes that his aim is not, in that paper, to present an original interpretation, but rather to find links between different areas of Frege's work. This is an approach, he argues, that may

---

[1] Cf. Juliet Floyd's discussion of Frege's criticism of the *Tractatus* in Floyd (2011).

make it possible to find ways of interpreting debated passages that surpass, remove or move cemented dividing lines between the interpretative traditions. In a similar vein, we present here a series of possible interpretations of different parts of Frege's work for the purpose of bringing out a picture of how they are related. By focusing in particular on the two endpoints of the scale that we looked at earlier—exact or strict science at one end and *Dichtung* at the other—we hope to shed light on what lies in between, the connecting links between areas that we tend to conceive of as far apart in Frege's work.

Gottfried Gabriel's book *Zwischen Logik und Literatur: Erkenntnisformen von Dichtung, Philosophie und Wissenschaft* (1991) served as an inspiration for the choice of this volume's title. In his discussion of science and fiction in the essay in the present volume, Gabriel immediately turns to questions that are at the core of its overall topic. He uses the ancient controversy between emotivism and cognitivism concerning the function of poetry, as a framework for an examination of Frege's conception of poetry and fiction, and of the difference between scientific and fictional discourse. The distinctions drawn from Frege are central to Gabriel's critical exploration of the conception of meaning that underlies an emotivist approach to poetry. He then seeks to defend the cognitivist view of poetry on semantic and pragmatic grounds—with the aid of Frege's distinctions—while stressing that poetry also has emotive functions.

Frege's conception of logic and of logical laws lies at the center of the next two essays in this volume. Jan Harald Alnes argues that Frege's conception of science includes three features that form "the scientific picture": (1) a science is applicable to other sciences, or even to itself, (2) a science consists of a more or less rigid system of judgments, and (3) a science presupposes elucidations and illustrative examples. The scientific picture guides Frege in his philosophical and logical reflections, in Alnes' view. He moves on to explore Frege's claim that the axioms of *Begriffsschrift* and *Grundgesetze* are obvious and need no justification, and Frege's classification of judgements as "analytic" or "synthetic", "a priori" or "a posteriori". Alnes aims to show that the current tendency to ascribe a heavy metaphysics to Frege reveals misunderstandings of his conceptions of logic, of logical laws and of judgement. A further important and parallel aim of this paper is to clarify the relation between scientific research and philosophical investigations. Anssi Korhonen's essay raises the question of what relationship logic bears to thought according to Frege, and identifies two lines of interpretation in the current debate. According to the first one, Frege thinks that the laws of logic are normative and prescriptive in relation to thought, while the second is guided by the assumption that he considers adherence to the laws of logic constitutive of thought. Here, normativity and constitutivity appear to exclude one another, and Korhonen explores this interpretative difficulty. His discussion is driven forward by a comparison to Kant's distinction between different kinds of 'deductive' or justificatory strategies in the Transcendental Aesthetic and the Transcendental Analytic in the first *Critique*.

The next essay takes up some aspects of Frege's conception of arithmetic. Sören Stenlund discusses Wittgenstein's criticism of Frege's inability to see the "justified side of formalism". Indeed, Frege's early critique of Thomae's "formal standpoint" shows that he fails to see the operative aspect of mathematical symbolism stressed by Thomae. In fact, Thomae's "formalism", correctly understood as a way of emphasizing the difference between a sign and a symbol, undermines the ontological conception of arithmetic represented by Frege.

The distinction between fact and fiction is central to Martin Gustafsson's discussion of Frege's judgment stroke. Frege's use of a judgment stroke in his conceptual notation has been a matter of controversy, at least since Wittgenstein rejected it as "logically quite meaningless" in the *Tractatus*. Recent defenders of Frege include Tyler Burge, Nicolas Smith and Wolfgang Künne, whereas critics include William Taschek and Edward Kanterian. Against the background of these defenses and criticisms, Gustafsson argues that Frege faces a dilemma the two horns of which are related to his early and later conceptions of asserted content, respectively. On the one hand, if content is thought of as something that has propositional structure, then the judgment stroke is superfluous. On the other hand, if what is to the right of the judgment stroke is conceived as a sort of name designating a truth-value, then there is no consistent way to avoid construing the judgment stroke as a kind of predicate, and thereby fail to do justice to the act-character of judgment and assertion.

In Gisela Bengtsson's essay, an assumption about Frege's *Begriffsschrift* that is at play in anti-semantic interpretative approaches is identified: the notion that translatability to Frege's concept script can serve as a criterion for deciding whether a thought is expressed by a sentence or utterance or not. The viability of such a translatability criterion is questioned by way of an examination of Frege's account of the aim and character of his logical language in relation to natural languages, and an investigation of his remarks on poetry, literature and fiction. Bengtsson argues that since sentences used in the context of *Dichtung* may have sense and be possible to understand, in Frege's view, the translatability criterion stands out as flawed. Frege's appeal to "an approach of willingness to understand" when seeking to shed light on the distinction of the logical language is brought into focus in this chapter, as is the relation between Frege's use of elucidations and his conception of *Dichtung*.

In the next chapter, Todor Polimenov examines Frege's discussions of fictional discourse. Polimenov argues that Frege sometimes appeals to fiction to get a background against which to draw the semantic boundaries of his logical investigations. Frege's use of examples from fiction to clarify some specific relations between his semantic concepts is worth analyzing, according to Polimenov. The remarks on fiction may contain insights, he suggests, that let us elaborate a Fregean definition of fictional discourse. Polimenov wants to show that these remarks do not just negatively say what fictional discourse is not, but also indicate what it is. Furthermore, he claims that it is important to distinguish between semantic and pragmatic features of Frege's view of fiction. The pragmatic ones, he argues, anticipate some basic insights of a speech-act theoretical approach to fictional discourse.

The volume's two final chapters focus on the relation between Frege and Wittgenstein. Allan Janik, in his discussion of Wittgenstein's legacy from Frege, places Frege in a philosophical context that is not ordinarily considered. The starting point is Wittgenstein's remark "Philosophie dürfte man eigentlich nur dichten"

(Wittgenstein 1998, 28) and the differing English translations of it. Janik argues that we must turn to Frege to find an understanding of Wittgenstein's view of the relation between philosophy and *Dichtung* and to see clearly the origin of the philosophical method of clarification used in Wittgenstein's later philosophy. In the final essay, Nuno Venturinha also examines Wittgenstein's appeal to *Dichtung*. His paper begins by scrutinizing classic approaches to the question of agrammaticality, with a particular focus on Frege and the early Wittgenstein. He then puts focus on the later Wittgenstein's treatment of nonsense-poems and claims that the failure of the *Philosophical Investigations* as a book is actually connected with Wittgenstein's recognition that philosophy should be written "under the form of poetry".

One of the aims of this volume is to explore Frege's conception of philosophy by looking at his way of characterizing and elaborating the relation between science and *Dichtung*. To ask questions about this relation is also to ask about the distinction between fact and fiction, and about what is found between them—such as philosophy, the humanities, the language of life. It is further to ask about the relation between thought and world, and between language and logic, according to Frege. The possibility of translation between languages, and the notion of communicating the same meaning in different languages, is hence a topic that becomes central to more than one of the essays in the volume, regardless of whether the focus of is on science or *Dichtung*. The diversity of focal points and interpretations brought out in the essays of this volume will hopefully bring out a more nuanced picture of Frege the philosopher who is often taken to subscribe to scientism. It also reflects the difficulties involved in placing Frege with regard to the distinction between a scientific and a philosophical approach that von Wright made use of in his outline of the origins of analytic philosophy.

References

Alnes, J. H. (2013). Trekk ved Freges logisisme: Begrep, logisk objekt og Aksiom V. *Norsk Filosofisk Tidsskrift, 48*(3–4), 243–264.
Black, M. (1939). Relations between logical positivism and the Cambridge school of analysis. *The Journal of Unified Science (Erkenntnis), 8*(1/3), 24–35.
Floyd, J. (2011). The Frege–Wittgenstein correspondence: Some interpretative themes. In E. de Pellegrin (Ed.), *Interactive Wittgenstein: Essays in memory of George Henrik von Wright* (pp. 60–75). Dordrecht: Springer Netherlands.
Frege, G. (1918). Der Gedanke: Eine logische Untersuchung. *Beiträge zur Philosophie des Deutschen Idealismus, I*, 58–77. English edition: Frege, G. (1997). Thought. In G. Frege, *The Frege reader* (pp. 325–345). M. Beaney (Ed.). Oxford: Basil Blackwell.
Gabriel, G. (1991). *Zwischen Logik und Literatur: Erkenntnisformen von Dichtung, Philosophie und Wissenschaft*. Stuttgart: Metzler.
von Wright, G. H. (1993). *Logik, filosofi, språk: Strömningar och gestalter i modern filosofi*. Nora: Bokförlaget Nya Doxa.
von Wright, G. H. (1994). Analytische Philosophie—eine historisch-kritische Betrachtung. In G. Meggle & U. Wessels (Eds.), *Analyomen 1: Proceedings of the 1st conference 'Perspectives in Analytical Philosophy'* (pp. 3–30). Berlin: Walter de Gruyter.

Waismann, F. (1940). Was ist logische Analyse? *The Journal of Unified Science (Erkenntnis)*, *8*(5/6), 265–289.
Weiner, J. (2001). Theory and elucidation: The end of the age of innocence. In J. Floyd & S. Shieh (Eds.), *Future pasts: The analytic tradition in twentieth century philosophy* (pp. 43–66). New York: Oxford University Press.
Wittgenstein, L. (1981). *Zettel* (2nd ed.). G. E. M. Anscombe & G. H. von Wright (Eds.). Oxford: Basil Blackwell.
Wittgenstein, L. (1998). *Culture and value/Vermischte Bemerkungen*. G. H. von Wright (Ed.). Collaboration with Nyman, H. (Rev. ed. by Pichler, A.; P. Winch, Trans.). Oxford: Basil Blackwell.

Chapter 2
Science and Fiction: A Fregean Approach

Gottfried Gabriel

Abstract In Frege's analysis of the relationship between science and fiction there are two important aspects, which the paper will discuss. It shows that Frege makes a strict distinction between *Dichtung und Wissenschaft* on the level of object language but not on the level of metalanguage. (1) In his "On Sense and Reference" and in scattered remarks elsewhere Frege explains the semantics (and pragmatics) of scientific and everyday discourse. As a kind of side product he presents an explication of the concept of fictional discourse concerning questions of illocutionary force and reference. Here Frege anticipates J. R. Searle's speech-act-theory of fictional discourse, which allows to understand works of fiction as consisting (at least partly) of fictional discourse. On the basis of Frege's distinctions this approach is defended against ontological arguments, which make use of terms like 'fictive entities' or 'non-existent objects' in the Meinongian tradition. (2) Frege excludes the connotative or figurative elements of language, called "colourings and shadings" (*Färbungen und Beleuchtungen*) of sense or thought, from the scientific use of language and assigns such elements to "the art of poetry" or "eloquence". The fact that the expression 'colouring' is itself a figurative term, raises a paradoxical question: To what extent does understanding Frege's own explanation of the difference between sense and colouring depend on the poetic or at least rhetoric use of language? There are reasons to believe that Frege was—even if only reluctantly—aware of this paradox. Otherwise he would not have repeatedly emphasized that explanations of categorial logical distinctions (particularly such as those between 'function' and 'object') cannot dispense with "figurative expressions" (like 'unsaturated' and 'saturated'). Insofar as such distinctions are "reliant upon the accommodating understanding of the reader", they pay tribute to a rhetoric of cognition. While Frege denies that colourings contribute to cognitive *content*, he must still admit that they make an indispensable protreptic contribution to *conveying* cognition.

Keywords Cognitivism • Emotivism • Fiction • Figurative language • Frege • Science

G. Gabriel (✉)
The Friedrich Schiller University, Jena, Germany
e-mail: gottfried.gabriel@uni-jena.de

There is an old controversy between so-called emotivism and cognitivism, especially in ethics and aesthetics, concerning the status of moral and aesthetic value-judgements. A similar controversy is to be found in the theory of poetry, only similar because this controversy does not concern the nature of judgements, i.e., the judgement of literary critics about works of poetry, but the nature of poetry itself. The question is, what function or purpose does poetry fulfil, or to put it more cautiously, what functions or purposes (in the plural) can poetry fulfil. Now, this question is not quite a new one, in fact it is as old as poetry itself, and this also applies to how to answer it. Since the days of Plato and Aristotle this question has been answered by comparing poetry with philosophy and science. Here, very often the position of poetry (and arts in general) has been a defensive one; poetry has been defended or justified as a subject in its own right against the claims of philosophy and science. An early example in the English tradition is the famous *Defence of Poetry* by Sir Philip Sidney (1971).

From the beginning, poetry was pushed into a defensive position by the question of whether and how it may convey truth and knowledge. So, defending poetry often consisted of showing that there was a place left in the field of knowledge, which was not yet occupied by science and philosophy. Seen historically, the most recent additions to this field were the mental and the social fields of study. Eighteenth century authors like Henry Fielding and Laurence Sterne are praised for their psychological observation and their analysis of human nature, and nineteenth century authors like Émile Zola are praised for their sociological insight and analysis of human society. Comparing poetry with sciences in this way became more difficult in the twentieth century, as there were now special sciences concerned with the mental and the social, namely psychology and sociology. When it came to the themes of literature, all the different facets of the human condition, which the field of knowledge is concerned with, were already distributed. For poetry the consequence seemed to be its end as an "organon of truth" (to use Hegel's formulation); also it appeared that poetry would have to give up claiming its own truth and would be reduced to repeating the truth of science for non-scientific people or that it would have to leave the field of truth and look for a different field of activity in which to sustain a relevant function. I think it is this alternative, which gives the emotive theory of literature its plausibility. As Frege put it: the function of literature is not to convey truth and knowledge but to affect our feelings (Frege 1983, 151f., 1997, 238ff., cf. 1918, 63, 1997, 330f.). Now, Frege's emotivism in the field of aesthetics and the theory of poetry is founded on his philosophy of language, especially on his concept of meaning. In this respect he is the father of modern emotivism in the analytic tradition (R. Carnap, I. A. Richards etc.).

What I want to do is to investigate the emotivistic approach by considering its concept of meaning. In fact, I want to argue in favour of the cognitivistic view of poetry on semantic and pragmatic grounds. The position I hold does not deny emotive functions of poetry but maintains (on the other hand) that poetry (at least some poetry) has cognitive value, i.e., may convey knowledge, as well. I would like to explicate the meaning *of* literature by analysing the concept of meaning *in* literature.

More generally speaking, poetry is something like a paradigm case in the field of the philosophy of language. The adequacy of a theory of language is sometimes even tested by showing its adequacy to poetry (*Dichtung*), especially to fiction. We can already find this proceeding in Frege. Explaining the semantics and pragmatics of scientific and everyday discourse in his *Über Sinn und Bedeutung* Frege presents as a kind of side product an approach to the semantics and pragmatics of fictional discourse and fiction as well. Though I do not agree with Frege's emotivistic conclusion concerning the function of fiction itself, his view of fictional discourse seems convincing to me.

To speak about fiction presupposes accepting factual reality as its counterpart. In contrast to panfictionalism, which attempts to annihilate the distinction between fact and fiction, we have to keep up the opposition between aesthetical 'illusion' and extra-aesthetical 'existence'. The confusion of fiction and reality, or rather the blurring of the boundary between fact and fiction, has itself often been made the topic of fictional literature (as in Cervantes' *Don Quijote*); the playful manipulation of this boundary is known as 'fictional irony'; and 'life is a dream' may be regarded a literary topos (e. g. with Calderón, Hugo von Hofmannsthal, Lewis Carroll). But we must keep in mind that all this is to be taken as a literary re-presentation (*Vergegenwärtigung*) of this confusion and its implications for life, not as the claim or assertion that there is no difference between fact and fiction.

The category of reality remains regulative in force even if, in some cases, it may be difficult or even practically impossible to distinguish between fact and fiction. In Frege the acknowledgment of reality corresponds to the presupposition that we refer to objects in assertoric discourse:

> Now we can of course be mistaken in the presupposition, and such mistakes have indeed occurred. But the question whether the presupposition is perhaps always mistaken need not be answered here; in order to justify speaking of the reference of a sign, it is enough at first, to point out our intention in speaking or thinking. (Frege 1892a, 31f., 1997, 156)[1]

What is important is that there are criteria for the distinction between fact and fiction, not that they will lead to a decision in every single case. It may even happen that these criteria become themselves subject to debate, but it is not possible to dispense with such criteria altogether: The distinction between fact and fiction and accordingly between science and poetry is a transcendental condition, that is, a condition of the possibility of every orientation in the world.

In scattered remarks in his papers Frege presents an explication of the concept of fictional discourse concerning questions of reference and illocutionary force. Here Frege anticipates J. R. Searle's (1975) speech-act-analysis of fictional discourse, which allows to understand works of fiction as consisting (at least partly) of fictional discourse.[2] On the basis of Frege's distinctions this approach can be defended

---

[1] The page numbers of Frege's works refer to the first German editions (cf. the bibliography). The English translation given here follows for the greater part the edition of Beaney (Frege 1997), which includes the original pagination.

[2] A similar approach based on Fregean distinctions is presented in Gabriel (1975). Cf. the short English version Gabriel (1979).

against ontological arguments, which make use of terms like 'fictive entities' or 'non-existent objects' in the Meinongian tradition.

Frege is mainly engaged in distinguishing science from fiction. The differentiation is carried out in three ways. Two of them are semantical and one pragmatical. The main distinction in Frege's semantic is that between sense (*Sinn*), reference (*Bedeutung*) and colourings (*Färbungen*). We can read and understand fiction. So fiction has sense and in understanding it we grasp the sense of its words and sentences, but in distinction to science, which claims reference, fiction lacks reference:

> Of course in fiction words only have a sense, but in science and wherever we are concerned about truth, we are not prepared to rest content with the sense, we also attach a reference to proper names and concept words; and if through some oversight, say, we fail to do this, then we are making a mistake that can easily vitiate our thinking. (Frege 1983, 128, 1997, 173)

Ordinary assertoric discourse and especially scientific discourse strives for truth and a necessary condition for sentences to be true (or to be false) is that its words have reference. Sentences in fiction need not be true but may be true. In that sense Frege's statement that "in fiction words only have a sense" is not quite correct. Sense is *sufficient*, but it is not excluded that words and sentences in fiction do have a reference, only they do not need to have one. To have reference is "a matter of no concern to us" in fiction (Frege 1892a, 33, 1997, 157). In the case that words, especially proper names (including definite descriptions) in fiction are without reference, the sentences are neither true nor false. The thoughts which are expressed by such sentences Frege (1983, 141f., 1997, 230) calls "mock thoughts" (*Scheingedanken*). This characterisation is a negative semantical one. If the words in fiction have reference the sentences are true or false, yet in both cases the poet does not perform the speech act of assertion. His utterances are without illocutionary force, especially without "assertoric force" (*behauptende Kraft*) (Frege 1918, 63, 1997, 330). This characterisation is a pragmatic one. It means, that the thoughts, which are presented in fiction, are presented without claiming their truth. Following Frege fiction is a deviant kind of discourse, namely a non-assertoric discourse without claims to reference and truth.

The kind of speech-act-analysis of fiction outlined above has been criticised for degrading the significance of poetry. The course of the mimesis-debate appears to feed this fear. Since Plato, poetry has had to justify itself in the face of philosophy and science, just as rhetoric has had to do in the face of logic. Distinctions such as those between 'proper' and 'improper' speech seem to confirm the suspicion that fiction has a deficient status compared with fact. Only in this way can we explain why attempts based on the theory of speech acts to define fictional discourse through its deviations from ordinary discourse have been misunderstood as exclusionary. J. R. Searle's (1979, 67) characterisation of fictional discourse as "parasitic" may have encouraged this misunderstanding. This is so because Searle goes no further than a negative characterisation, without treating in any detailed fashion the positive possibilities of fictional literature. The self-limitation of a single author, however, cannot be regarded as the failure of the theory of speech acts in its entirety. In any event, the explication of fictional discourse in terms of a theory of deviance implies neither the exclusion of fictional discourse nor a hierarchical order in which assertoric discourse is placed above fictional discourse. The negative characterisation only states the obligations from which fictional discourse has been freed to be able to fulfil its peculiar function in the form of poetry, that is, as fictional literature, complementary to other forms of cognition. The point of such analysis is precisely to make clear in comparison, for example, with history, that poetry, despite its fictionality, can have cognitive value. This is not the position of Frege, who defends an emotivistic view of fiction, but at least there is a Fregean way to justify a cognitivistic position.

Before we later go over to defend the cognitive value of fiction, we have to distinguish fictional discourse from discourse about fiction. Whereas the first is non-assertoric the second is assertoric. The question therefore is, what are the assertions about in this case, i.e., what is the reference of fictional proper names (or other expressions which appear to refer to fictive objects like, for instance, definite descriptions). One answer, which was given by A. Meinong and some Neo-Meinongians (cf. especially Parsons 1980, 1982), is that such discourse is about fictive, i.e., nonexistent objects. (I distinguish 'fictive', a predicate applying to objects, from 'fictional', a predicate applying to texts, stories, discourse, and so on). The next question is, whether we have to accept fictive objects. Frege himself did not comment on their possibility. In spite of that his distinction between customary (*gewöhnlicher*) and indirect reference (*ungerader Bedeutung*) helps us to eliminate talk about fictive objects in favour of talk about fictional texts etc.

The punchline of Frege's distinction is that the indirect reference of a word is its customary sense. Let's consider a simple example: The sentence 'Little Red Riding Hood gathered flowers' does not occur only in the fairy-tale *Little Red Riding Hood*. It might be an answer to a question like 'What did Little-Red-Riding-Hood do when she left the path?' as well. In this case the sentence is not used in a fictional but in an assertoric way. To make clear that the assertoric utterance of the sentence 'Little Red Riding Hood gathered flowers' is not about a fictive object called 'Little Red Riding Hood' we ask for its truth-conditions. The assertion is true if and only if it is true, that the fairy-tale *Little Red Riding Hood* tells us (says so) that Little Red Riding Hood gathered flowers. Now the truth of the utterance is a question of the *sense* of the text of the brothers Grimm. The descriptive name 'Little Red Riding Hood' here occurs in indirect reference (i.e., in an intensional context) and therefore, as Frege pointed out, its reference is a sense and not an object. In this way the utterance is analysed as an assertion not about a fictive object but about the content of a fictional text. For the analysis of more complicated cases cf. Gabriel (1993). Frege seems to treat his own example sentence 'Odysseus was set ashore at Ithaca while sound asleep' only as a possibly fictional one, but it might be used assertoric as well, namely in an assertion about the content of Homer's *Odyssey* (1892a, 32, 1997, 157).

Critics might argue that we do avoid acknowledging fictive objects but only at the cost of acknowledging intensional ones. People understand Frege's idea, that the indirect reference of expressions is their customary sense, in such a way that the senses coordinated with names (proper names and definite descriptions) become intensional objects in indirect speech. Although Frege sometimes suggests this interpretation, it is still a step in the wrong direction. The distinction between sense and reference is a distinction between semantic roles and semantic roles should not be confused with ontological objects. From the fact that people speak of senses in indirect speech, one cannot without further argument conclude that these senses are ontological objects of a special kind. Now in order to avoid acknowledging intensional objects, one can analyse references to senses as references to meaningful uses of linguistic expressions. Within these limitations, fictive objects like literary figures can be conceived of as individual senses constituted by the corresponding fictional texts.

The arguments given above consist of two parts. First, I have tried to show that so-called fictive objects in so far as they are needed at all can be captured as Fregean senses. Second, I added an interpretation of senses, which avoided conceiving them as intensional objects in the ontological sense. Intensions are linguistic, not metaphysical objects. Hence the outlined approach is a Frege-Wittgensteinian one. Compare Gabriel (1975, 33–42, 1979, 249–253). It is possible both to accept the first part without accepting the second and to accept the second without the first. It is therefore not surprising that, independently of its application to problems concerning the semantics of fiction, the non-ontological conception of senses (intensions) is relevant to such traditional philosophical discussions as the nominalism-Platonism debate. The position defended here can be characterised as non-psychological conceptualism.

Frege's understanding of science becomes especially clear in his view of logic and the role which logic plays in science. Logic, with its orientation on the concept, neglects the sensuous side of thought and eliminates the "intuitive" (Frege 1879, IV, 1997, 48). Exemplary are Frege's efforts to keep the 'purely' logical free of all sensuous-psychological 'impurities'. In the philosophy of language, this purism finds its expression in the disregard of those elements of language whose contribution to the semantic content of propositions is irrelevant in terms of truth-value and to this extent of no relevance for making inferences. Among these elements are grammatical and pragmatical aspects of language that have to do with the relation between the speaker and the listener:

> Now all those features of language that result only from the interaction of speaker and listener—where the speaker, for example, takes the listener's expectations into account and seeks to put them on the right track even before a sentence is finished—have no counterpart in my formula language, since here the only thing that is relevant in a judgement is that which influences its *possible consequences*. (Frege 1879, 3, 1997, 54)

In his conception of science Frege connects his *Begriffsschrift* with the ideal of a system, explicitly formulated in the posthumous paper *Über Logik in der Mathematik* (Frege 1983, 221, 1997, 310), but conceived already in *Die Grundlagen der Arithmetik* (Frege 1884, § 3, 1997, 93). Here Frege gives an interpretation of the concepts 'analytic', 'synthetic', 'a priori' and 'a posteriori' in such a way that these become proof theoretic meta-predicates for judgements and for sciences as well. (Of course proof theory is not to be understood here in the sense of Hilbert. Hilbert's conception is criticised by Frege). By tracing the chains of inference from the judgements of a science backwards to its premises, it is Frege's ideal to find out the totality of axioms, basic laws or laws of nature, from which the judgements of this science are deducible by purely logical inferences. This ideal does not correspond to the logic of discovery (*ars inveniendi*) but to the logic of justification (*ars iudicandi*). Frege is not interested in the genesis but only in the validity of scientific knowledge. Looking at the foundation of a science he gives—in dependence of the provability of the particular judgements—a classification of sciences, which we can put in the following way:

(a) A science is *analytic* if and only if all proofs are reducible to logical basic laws and definitions.
(b) A science is *synthetic* if and only if there is among its basic laws at least one synthetic law.
(c) A science is *a priori* if and only if all its basic laws are a priori.
(d) A science is *a posteriori* if and only if there is among its basic laws at least one a posteriori law. Frege (1884, § 3, 1997, 93) explains: "For a truth to be *a posteriori*, it must be impossible for its proof to avoid appeal to facts, that is, to improvable and non-general truths that contain assertions about particular objects."
(e) A science is *synthetic a priori* if and only if all its basic laws are a priori, and there is among these basic laws at least one synthetic law.

The ambition of Frege's logicism was to show that arithmetic is a purely analytic science. Because of Russell's antinomy he gave up this program but he continued to transform modes of inference, which do not belong to logic, like the inference by mathematical induction from n to n + 1, into axioms. This is the case in the system of the Peano axioms. So the epistemological nature of arithmetic is still determined by the epistemological nature of the axioms (Frege 1983: 219f., 1997, 308f.).

The sketched classification of sciences is of course only possible for such sciences, which allow an axiomatic-deductive representation, and especially not for the humanities. Accordingly Frege underlines:

> What are called the humanities (*Geisteswissenschaften*) are closer to poetry (*Dichtung*), and are therefore less scientific, than the exact science, which are drier in proportion to being more exact; for exact science is directed toward truth and truth alone. (Frege 1918, 63, 1997, 330)

What is excluded in Freges view of logic and science (following the quotation from the *Begriffsschrift*) is, for example, the subject/predicate distinction, which is replaced by the argument/function distinction. Also excluded from the scientific use of language are the connotative or figurative elements of language, called "colouring and shading" (*Färbungen und Beleuchtungen*) of sense and thoughts (Frege 1892a, 31, 1997, 155). Frege goes so far and too far (cf. Gabriel 1975, 125) when he declares:

> It makes no difference to the thought whether I use the word 'horse' or 'steed' or 'nag' or 'prad'. The assertoric force does not cover the ways in which these words differ. What is called mood, atmosphere, shading in a poem, what is portrayed by intonation and rhythm, does not belong to the thought. (Frege 1918, 63, 1997, 331)

Freges German examples are '*Pferd*', '*Roß*', '*Gaul*' and '*Mähre*'. Other examples are 'walk (*gehen*)', 'stroll (*schreiten*)', 'saunter (*wandeln*)' and 'dog (*Hund*)', 'cur (*Köter*)' (Frege 1983, 152, 1997, 240f.) For Frege colouring and shading belong to the realm of mental "ideas" (*Vorstellungen*). In contrast to sense and reference they are therefore not objective but subjective "and must be evoked by each hearer or reader according to the hints of the poet or the speaker" (Frege 1892a, 31, 1997, 155).

Whereas for Frege such elements ought to be eliminated from science they are essential for poetic fiction. So the two negative characterisations of *Dichtung* already explicated are supplemented by a positive surplus of these "colourings". Frege explicitly assigns such elements to "the art of poetry" or "eloquence" and hereby specifies semantical elements, which transform fictional discourse into fictional literature.

The Fregean expression 'colourings' is a translation of the Latin term '*colores*' of ancient rhetoric, in which the function of winning over listeners through colourings of the matter to be represented is attributed to the ornamentation (*ornatus*) of speech (Quintilianus 1995, book IV, Chap. 2, §§ 88–100). With regard to ancient usage in general cf. Ernesti (1962a, 383–385, 1962b, 63–66). The terminology found acceptance in poetics (Scaliger 1987, 121). Frege's form of expression coincides remarkably with that of Ernesti (1962a, 384) in his German commentaries on ancient rhetoric. The latter speaks like Frege of "*Beleuchtung*" along with "*Colorit*". Most of all, he defines *Colorit* as "character of expression with respect to *sense and thought*" (emphasis G. G.).

One still speaks today of favourable or biased reports and presentations as 'coloured'. Antiquity itself, however, presents a neutral view with respect to colourful embellishment or portrayal. Both usages can be found in Frege. In no way does he devalue colourings in general; instead, he only denies their cognitive value. He explains this denial by arguing that while colourings may indeed have an emotive effect they are in distinction to sense "not objective" cognitive contents (Frege 1892a, 31, 1997, 155f.). Here Frege is following a view found throughout the epistemological tradition. The corresponding demand placed on philosophical style is that it has to represent the "naked" truth, rejecting "the embellishment of words that aids the orator" (Wolff 1996, § 149f.). The metaphor that truth is supposed to appear "naked" should give the philosopher reason to think about his own rhetoric.

The fact that the expression 'colouring' is itself a figurative, metaphorical term, raises a paradoxical question: To what extent does understanding Frege's own explanation of the difference between sense and colouring depend on the poetic or at least rhetoric use of language? There are reasons to believe that Frege was aware of this paradox. So in his elucidation of the concept of thought in the late paper *Der Gedanke* he complains about the linguistic difficulties of the philosophical categorial discourse:

> Something in itself not perceptible by sense, the thought, is presented to the reader—and I must be content with that—wrapped up in a perceptible linguistic form. The pictorial aspect of language presents difficulties. The sensible always breaks in and makes expressions pictorial and so improper. (Frege 1918, 66, footnote, 1997, 334)

In his earlier paper *Ueber Begriff und Gegenstand* Frege even emphasises, that elucidations (*Erläuterungen*) of categorial logical distinctions—particularly such as those between 'function', 'concept' and 'object'—cannot dispense with figurative expressions like 'unsaturated', 'complete' and 'saturated'. These expressions "are of course only figures of speech; but all that I wish or *am able* to do here is to give hints (*Winke*)" (Frege 1892b, 205, 1997, 193; emphasis G. G). About these hints Frege (1904, 665) even says:

> I have to limit myself (*muß mich darauf beschränken*), to hint by a figurative expression at that what I mean, and I am reliant upon the accommodating understanding of the reader.

The categorial elucidations belong to the level of the logical metalanguage, and they pay tribute to a rhetoric of cognition. While Frege denies that figurative and metaphorical colourings contribute to cognitive content, he admits that they might give hints for a better understanding and make an indispensable protreptic contribution to conveying categorial cognition: "Where the main thing [...] cannot be conceptually grasped, these constituents are fully justified" (Frege 1918, 63, 1997, 330 f). In this case connotative elements of language obviously *do* have cognitive value—even in logic. Cf. the extensive treatment of this matter in Gabriel (1991, 65–88, especially 79 ff.) and also Schildknecht (2002, part 2.)

We can summarise: In his analysis of the relationship between science and fiction Frege makes a strict distinction between *Wissenschaft* and *Dichtung* on the level of the object language but not on the level of the philosophical metalanguage. Here he goes back to semantical elements, which he ascribes to poetry otherwise. Frege does not go as far as Wittgenstein (1980, 24) who declares: "Philosophy ought really to be written only as a *poetic composition*" (*Philosophie dürfte man eigentlich nur* dichten). But he concedes—even if only reluctantly—that at least sometimes the categorial discourse of philosophy is forced to use poetic language. This result implies, that colouring—in contrast to Frege's own view—might perform a cognitive contribution. Frege's emotivistic conception of fiction, according to which fiction has the function of evoking the emotions of the reader, is therefore to be supplemented by a cognitivistic approach. Using Fregean distinctions such an approach will now be developed systematically.

We understand the meaning of every day and scientific sentences when we know what they are saying, when we understand the sense and know the reference of their words. (Here we are dealing with declarative sentences only). It is the case, however, that sentences may have meaning not only by saying something. Generally speaking one can mean something by description or by re-presentation (*Vergegenwärtigung*). Sentences which report actions, events, situations etc., as if they had taken place, can, in the second sense then, have meaning, if what they pretend to report can be understood as standing for something else. This kind of meaning relation may be called the relation of the particular to the general. The difference in question can be formulated as follows: in the first case a sentence merely means what it says, and in the second it also means what it shows. A sentence or a text can consequently mean more than it says. Especially poetic texts mean more than they say. They may do so by suggestion, connotation, contextual implication, etc. A poetic or literary work of fiction then is a text, which means more than it fictionally—in fictional discourse—says. Here Freges colourings play a central role. Semantically, what we have here is a new direction of meaning, a transition from referential meaning to symbolic meaning. The category of the particular is central to an identification of the cognitive value of literature, at least as long as we avoid G. Lukács' (1967) limited view. Together with the concept of changing the direction of meaning, it makes up the common core of the various aesthetic conceptions of Baumgarten's "*perceptio praegnans*", Kant's "aesthetic idea", Goethe's "symbol", Cassirer's "symbolic pregnancy", and Goodman's "(metaphorical) exemplification". Following Kant's *Kritik der Urteilskraft* (Kant 1968, § 49) it is impossible to explicate an aesthetic idea completely by conceptual thinking. If we construe this insight semantically, we can identify an aesthetic idea of a poetic work with the surplus of meaning, i.e., the whole of all the possible connotations implied contextually by the text of the work. Connotations of this kind are not specific to literary works of fiction alone; they are essential to literary works of non-fiction as well.

A literary work of fiction may be true even though it contains no true statements. The truths *of* fiction must not be reduced to truths (true statements) *in* fiction. They are truths, which are not *told in* but *shown by* the text; they are truths to which the recounted events, persons, and things are in the relation of the particular to the general. Interpretation of a literary work of fiction has then to find out the undetermined general to a given particular, which is the whole of the recounted events, persons, and things.

Admittedly, fictional literature does also convey true propositional information without asserting it. In Frege's words:

> In poetry we have the case of thoughts being expressed without being actually put forward as true, in spite of the assertoric form of the sentence; although *the poem may suggest to the hearer that he himself should make an assenting judgement*. (Frege 1918, 63., 1997, 330; emphasis G. G.)

But what is the function of true information within a work of fiction? To answer this question, let us look to an example, namely at the initial passage of Gottfried Keller's novel *Der grüne Heinrich* (*Green Henry*) in its first edition:

> Among the most beautiful cities, especially in Switzerland, are those which lie at the same time by a lake and a river, so that, like a wide gateway at the end of the lake, they immediately absorb the river, which runs right through them into the country. Such as Zurich, Lucerne, Geneva; and Constance, too, is in a way one of them. (Keller 1956, 9)

The place names mentioned here refer to real cities—it is possible to relate them to actually existing objects. The passage quoted could almost have been taken from a travel guide, for instance one advertising trips to Constance. In the context at hand, they induce the reader to take an imaginary boat trip: "It is hard to imagine anything more pleasant than a tour on one of these lakes, for example the Zurich one. Board the ship to Rapperswyl [...]." In the course of the imaginary boat trip, which then follows, Zurich and its surroundings are descriptively made present in their historical and geographic characteristics (for about two pages in the novel). Although composed in a poetically rich language, the presentation does (if related to the time in which the story is set) not only make reference to real places, but is moreover verifiable (the propositions stated are true). It could therefore well be part of a classical travel account, which is not only concerned with providing a route for travelling, but also tries to capture the atmosphere of things. However, the passage is in fact part of a novel, and this determines our attitude towards it. With a travel account, we may delight in its poetic qualities, but we will insist on its reference and truth. In a realist novel, by contrast, reference to actually existing entities and truth constitute a frame for the actions of the characters (in accordance with the demand for realistic 'plausibility'); but we do not expect that they have to be true in every detail. It usually suffices if the novel is consistent with well-known facts, if it does justice to them; but we do not insist on the justification of these facts. Thus, while propositional truths may play a vital role in such cases (as background knowledge), they are not being *asserted* in a strict sense. We do not expect that the claim to truth is pursued. Rather, we expect that the poetic re-presentation will surpass factual truth by endowing it with a symbolic meaning for the situation of the characters within the novel. For instance, the juxtaposition of lake and river becomes a symbol of the contrast between rest (systole) and motion (diastole) in the life of Green Henry. In his novel, the author Keller comments directly on the change of direction in meaning. Following the description of the boat trip, the text takes up the motif from the beginning as follows:

> And so, the charm of the location near lake and river is similar for Lucerne and Geneva, and yet at the same time quite different and peculiar to each. To add to those cities an imaginary one in order to plant there the green seed of poetry may be in order. After having established a *sense of reality* through existing examples, fancy claims its place again; and all misinterpretation is prevented. (Keller 1956, 11, my emphasis. This refers to what I have called 'plausibility' above)

"Misinterpretation" is here to be taken as a reading which attempts to establish reference to reality. Concerning the second edition of *Green Henry*, Keller notes that he has filled the book with "all kinds of fibs" in order to "make it more clearly a novel". He further explains that there are still "donkeys", who "take it at autobiographical face value" (Keller 1956, 1155, letter to Maria Mellos 29.12.1880). It is granted that *some* biographical elements are incorporated in the novel. However, in a literary treatment of personal experience, an author will have to express these experiences for others in an exemplary way; he will have to turn the singular into a particular, thus abandoning its referential status.

In general, we can say the following: realistic, naturalistic, and historical novels must, to a certain extent, refer and be verifiable; however, this is not the basis of the specific cognitive value of such texts. Even in cases where fictional literary texts lay a claim to conveying *new* propositional knowledge—such as insights into the reality of life—the cognitive achievement does not consist in assertions of abstract general propositions. Rather, it consists in concrete non-propositional, narrative re-presentation of the content of such propositions. The first sentence of Tolstoy's novel *Anna Karenina* serves as a standard example for such an abstract general proposition in literary criticism: "All happy families are alike; each unhappy family is unhappy in its own way." It is clear that this sentence guides the reading of the novel, if only because of its prominent position. It would be absurd, however, to identify this sentence with the cognitive value of Tolstoy's novel. Asked about the "main thought" of his work, Tolstoy replies that in order to say what he intended to express he would have to write the same novel over again (Tolstoy 1978, 296f.). In other words, the cognitive achievements of narrative re-presentation cannot be captured exhaustively in a propositional way.

Anyway there might be general statements, implied contextually or shown by the text, which the author of a work of fiction wants to communicate as his message. Following Beardsley (1958, 403f., 409ff.) we may call such statements 'theses', but a thesis is not an argumentative speech act. Indeed, a sensible reader of a literary work of fiction will not expect that the truth of a thesis will be defended in the text. It may be defended outside the text by the author or the reader, but this is another question. Hence a thesis of a literary work of fiction is no assertion. On the other hand, the reader will expect that the author (a) himself believes that the thesis of his text is true, and therefore (b) accepts the consequences of the thesis.

I want to stress that literary works of fiction, which do not show or imply theses may nevertheless convey knowledge. They may fulfil the relation of the particular to the general in other ways than presenting a thesis. And this seems to be not the exception but the norm. When literature teaches us to see the world in a new way, or when it confirms our view of the world (which can also be a genuine gain in insight), this rarely happens in a propositional mode of speaking. The kind of knowledge we have to do with in these cases is knowledge by acquaintance. Through literature we become acquainted with situations, feelings, forms of life etc. more often than through being confronted with implied theses. The knowledge conveyed here cannot be propositional knowledge. The specific cognitive value of literature lies in its capacity to make things and situations present, i.e., to re-present them. This is the case even when the work of fiction contains or implies theses.

The cognition of things and situations cannot only be conveyed through propositional descriptions, but also through making situations present, thus becoming acquainted with them. Accordingly, we have to distinguish between the propositional knowledge *that* something is the case (knowledge by description) and the non-propositional knowledge *how* it is or would be to find oneself in a certain situation (knowledge by acquaintance).[3]

When the situations of others (including literary characters) are made present to us, this has a cognitive value because it can broaden the horizon of our understanding. It allows us to partake, in our imagination, of a wide range of different situations, motives, emotions, attitudes, perspectives, and sentiments, most of which we would never have come across in real life—or which we are spared.[4] What we have here is therefore not an immediate, 'actual' acquaintance; it is not a direct epistemic contact with things. Imaginary re-presentation does not aim at the "production of presence" (Gumbrecht 2004) in a real sense. Its aim is a cultivation of our reflective

---

[3] For a justification of the recognition of non-propositional cognition from the perspective of perception theory, cf. Schildknecht (2002, 199–215).
[4] For the role of imagination in the reception of fictional literature cf. Sutrop (2000).

judgement (in the Kantian sense). The successful literary presentation of attitudes or ways of life has a cognitive value independently of whether we approve of these attitudes or ways of life or abominate them. It shows us the *conditio humana* and, if need be, also its corruptions or perversions.

The idea that we partake in our imagination of the lives of literary figures, in the sense of *cognitive* 'empathy', is supported by a rehabilitation of emotions in current epistemology. Thus, already Robert Musil (1978, vol. 1, 494) emphasised that "intellect and emotions are not enemies". Two functions of emotive language have to be distinguished: its use to appeal and its use to make things present. The emotivistic view of literature holds that the function of literature is to convey emotions by evoking or awakening them. This causal way of putting things implies that it is vital to actually *feel* the emotions conveyed. But this is an undue restriction. The presentation of the feeling of alienation in Kafka's works does not primarily intend to evoke this feeling, but rather to make it comprehensible.

The preceding analyses should have made clear, why fictional literature can have a cognitive value despite (or rather by virtue of) its fictionality. In this respect, fictional literature often comes closer to real life than history precisely because its aesthetically complex, i.e., detailed and nuanced re-presentations do not depend on the truth of singular facts. Its cognitive achievement is to exemplarily make present the *conditio humana,* the human situation in the world. Thus, while fictional literature may be concerned with the same reality as science, it is not interested in mere facts, but rather in viewing reality from a human perspective.

Summarizing our results concerning meaning and cognition in science and fiction, we may put it in the following way. Instead of the traditional bipartite distinction between meaning as object and meaning as content, which also underlies the Fregean distinction between reference and sense, we have a tripartite distinction. The symbolic relations, which are concerned here become clearer by distinguishing between three corresponding kinds of meaning acts. First referring, second saying, and third showing (illocutionary acts are not considered here). Roughly speaking, in scientific texts and every day discourses, knowledge is conveyed by the acts of referring and saying, in literary fiction by saying and showing, and in non-fictional literature (as in the texts of Wittgenstein) by referring, saying, showing. The essential step in favour of the cognitive value of literature is to go beyond the propositional acts of referring and saying, and to acknowledge the act of showing as a meaning act in its own right.

References

Beardsley, M. C. (1958). *Aesthetics: Problems in the philosophy of criticism*. New York: Harcourt, Brace and Company.
Ernesti, J. C. T. (1962a). *Lexicon technologiae graecorum rhetoricae (1795)*. Hildesheim: Olms.
Ernesti, J. C. T. (1962b). *Lexicon technologiae latinorum rhetoricae (1797)*. Hildesheim: Olms.
Frege, G. (1879). *Begriffsschrift, eine der arithmetischen nachgebildete Formelsprache des reinen Denkens*. Halle: L. Nebert.

Frege, G. (1884). *Die Grundlagen der Arithmetik: Eine logisch mathematische Untersuchung über den Begriff der Zahl*. Breslau: W. Koebner.
Frege, G. (1892a). Über Sinn und Bedeutung. *Zeitschrift für Philosophie und Philosophische Kritik, 100*, 25–50.
Frege, G. (1892b). Ueber Begriff und Gegenstand. *Vierteljahrsschrift für wissenschaftliche Philosophie, 16*, 192–205.
Frege, G. (1904). Was ist eine Funktion? In S. Meyer (Ed.), *Festschrift Ludwig Boltzmann gewidmet zum sechzigsten Geburtstage, 20. Februar 1904* (pp. 656–666). Leipzig: J. A. Barth.
Frege, G. (1918). Der Gedanke: Eine logische Untersuchung. *Beiträge zur Philosophie des Deutschen Idealismus, I*(2), 58–77.
Frege, G. (1983). *Nachgelassene Schriften* (2nd ed.). H. Hermes, F. Kambartel, & F. Kaulbach, (Eds.). Hamburg: Felix Meiner.
Frege, G. (1997). *The Frege reader*. M. Beaney (Ed.). Oxford: Blackwell.
Gabriel, G. (1975). *Fiktion und Wahrheit: Eine semantische Theorie der Literatur*. Stuttgart-Bad Cannstatt: Frommann-Holzboog.
Gabriel, G. (1979). Fiction—A semantic approach. *Poetics, 8*, 245–255.
Gabriel, G. (1991). *Zwischen Logik und Literatur: Erkenntnisformen von Dichtung, Philosophie und Wissenschaft*. Stuttgart: J. B. Metzlersche Verlagsbuchhandlung.
Gabriel, G. (1993). Fictional objects? A 'Fregean' response to Terence Parsons. *Modern Logic, 3*, 367–375.
Gumbrecht, H. U. (2004). *Production of presence: What meaning cannot convey*. Stanford: Stanford University Press.
Kant, I. (1968). Kritik der Urteilskraft. In *Kants Werke: Akademie-Textausgabe* (Vol. 5). Berlin: Walter de Gruyter.
Keller, G. (1956). *Sämtliche Werke und ausgewählte Briefe* (Vol. 1). C. Heselhaus (Ed.). München: Hanser.
Lukács, G. (1967). *Über die Besonderheit als Kategorie des Ästhetischen*. Neuwied: Luchterhand.
Musil, R. (1978). *Briefe 1901–1942* (Vol. 2). A. Frisé (Ed.). Reinbek: Rowohlt Verlag.
Parsons, T. (1980). *Nonexistent objects*. New Haven: Yale University Press.
Parsons, T. (1982). Fregean theories of fictional objects. *Topoi, 1*(1-2), 81–87.
Quintilianus, M. F. (1995). *Institutio Oratoria: Ausbildung des Redners. Zwölf Bücher.* (3rd ed., H. Rahn, Trans., 2 vols). Darmstadt: Wissenschaftliche Buchgesellschaft.
Scaliger, J. C. (1987). *Poetices libri septem (1561)*. Stuttgart: Frommann-Holzboog.
Schildknecht, C. (2002). *Sense and self: Perspectives on nonpropositionality*. Paderborn: mentis.
Searle, J. R. (1975). The logical status of fictional discourse. *New Literary History, 6*, 319–332.
Searle, J. R. (1979). *Expression and meaning: Studies in the theory of speech acts*. Cambridge: Cambridge University Press.
Sidney, P. (1971). *A defence of poetry (1595)* (2nd ed.). J. A. van Dorsten (Ed.). Oxford: Oxford University Press.
Sutrop, M. (2000). *Fiction and imagination: The anthropological function of literature*. Paderborn: mentis.
Tolstoy, L. (1978). Letter to N. N. Strachov from the 23rd and 26th April 1876. In F. Christian (Ed.), *L. Tolstoy, Tolstoy's letters* (pp. 296–297). London: Athlone Press.
Wittgenstein, L. (1980). *Vermischte Bemerkungen/Culture and value* (2nd ed.). G. H. von Wright (Ed.). Oxford: Blackwell.
Wolff, C. H. (1996). *Discursus praeliminaris de philosophia in genere. Latin-German text*. G. Gawlick, & L. Kreimendahl (Eds.). Stuttgart-Bad Cannstatt: Frommann-Holzboog.

Chapter 3
Frege's Unquestioned Starting Point: Logic as Science

Jan Harald Alnes

Abstract Frege's conception of science includes three features: (1) a science is applicable to other sciences, or even to itself, (2) a science consists of a more or less rigid system of judgements and (3) a science presupposes elucidations, illustrative examples and a "catch on" among scientists. Together, I label these three features "The scientific Picture". Both logic and mathematics are included among the sciences and are covered by the scientific picture. As I understand Frege, this picture guides his logical and philosophical reflections. Here it is invoked in a treatment of two well-known and controversial Fregean topics: His claim, often repeated, that the axioms of *Begriffsschrift* and *Grundgesetze* are obvious and stand in no need of justification, and his use of a Kantian terminology in classifying judgements as analytic or synthetic, *a priori* or *a posteriori*. The most significant consequence of my reading is that it underscores the epistemological nature of Frege's thinking and, at the same time, downplays a current, and in my mind unfortunate, trend of ascribing to Frege a rather "thick" metaphysics. Towards the end, I discuss different aspects of the notion of a judgment at play in Frege's discussions: judgement as movement from thought to truth-value and judgement as represented by the judgement-stroke. These aspects point back to the distinction, so nicely illustrated by Frege's own writings, between a scientist, engaged in scientific research, and a philosopher, explicating the scientific activity and its general presuppositions, respectively.

Keywords Frege • Elucidation • Epistemology • Logic • Logicism • Metaphysics • Science

---

J.H. Alnes (✉)
UiT The Arctic University of Norway, Tromsø, Norway
e-mail: jan.harald.alnes@uit.no

© Springer International Publishing AG 2018
G. Bengtsson et al. (eds.), *New Essays on Frege*, Nordic Wittgenstein Studies 3,
https://doi.org/10.1007/978-3-319-71186-7_3

3.1 Introduction

My purpose here is to explore basic features of Frege's thinking in light of his general conception of science. Originally, the idea behind this approach towards Frege's writings was to get a fresh view on an old and much debated question among Frege-scholars, viz., the purpose and significance of his logicism. Some maintain that it is a strictly mathematical programme; others argue that it is philosophical as well, or even mainly philosophical. The latter reading divides into two branches; some take it to be an epistemological-metaphysical project and others interpret it as an epistemological project primarily. Robin Jeshion renewed the controversy in 2001 with an influential and controversial paper, arguing in favour of an epistemological-metaphysical account (Benacerraf 1981; Kitcher 1979; Jeshion 2001, 2004; Shapiro 2009; Bar-Elli 2010; Weiner 1990, 2004). In opposition to Jeshion, I maintain that Frege's logicism is an epistemological project. My principal reason is that Frege presents a general understanding of science, and that we should avoid ascribing to him substantial metaphysical assumptions or presuppositions that go beyond this understanding.[1] However, I shall set the numerous proposals in the secondary literature to the side. This, because my present topic goes beyond my original motivation for exploring Frege's logicism. I have come to think that Frege's conception of logic as science influences his reflections on all significant philosophical-logical questions. I shall attempt to make this view plausible.

The structure is as follows. In Part 2, I present three characteristics of Frege's conception of science and of logic as a science. The first concerns the relationship of the science of logic to the rest of the sciences. The universal applicability of logic makes it a distinguished member of the grand scientific enterprise. A major purpose of my discussion of the applicability of logic is to refute a reading of Frege, to the effect that he started out with a Kantian formal understanding of logic, but formed his mature conception when he began considering logical objects and logical singular terms.[2] The second characteristic is the all-present aim of scientific activity, namely, to make the advance from a judgeable content, as Frege would say before he introduced the sense-meaning distinction, or a thought, to a judgement. The third is that all sciences depend on elucidations of their basic notions. I next turn to the topics of logic as a systematic science and the role of philosophy, or, to be more precise, *Erkenntnislehre* or epistemology, for the classification of judgements as analytic, synthetic, *a priori* or *a posteriori*.[3] I argue that the topic of epistemology is not the act of judging or asserting, but the justification of a judgement or an assertion.

---

[1] In this, I agree with Joan Weiner. Like her, I take Fregean logicism to be a mathematical-philosophical project (*pace* Jeshion 2001, 940(n1); cf. Weiner 2004).

[2] That is to say, the claim is that his understanding of logic was transformed when he reached the second stage of the programme outlined in the preface to *Begriffsschrift*: "I sought first to reduce the concept of ordering-in-a-sequence to the notion of logical ordering, in order to advance from here to the concept of number" (Frege 1879a/1972, 104).

[3] By this, I do not mean that to Frege, epistemology covers all of philosophy, but only that I am focusing on the part of philosophy that concerns my present purpose.

I further maintain that in this restricted sense, epistemology plays a crucial role when Frege reflects on the status of the axioms of *Begriffsschrift* and *Grundgesetze der Arithmetik*. These reflections are not of a scientific nature, but take scientific judgements as their base. It follows from my discussion that the complex notions of "self-evident" and "obvious" employed by the aforementioned epistemological-metaphysical understanding of Frege's logicism are foreign to him. The topic of the fourth and final part is the dual role of Frege's notion of a judgement: its use in characterising the aim of science, and its role within the logical system of *Begriffsschrift* and *Grundgesetze,* respectively.

3.2 Science and Logic

One of the most impressive features of the sciences is their applicability: not only to everyday tasks of all kinds, but also to each other. At the outset of his career as a logician, Frege singles out the applicability of the new logical notation (hereafter called "Begriffsschrift") to the other sciences as one of its main merits.[4] In the 5 years between the publications of *Begriffsschrift* (1879) and *Die Grundlagen der Arithmetik* (1884), Frege wrote a series of articles—some published, others rejected by the publishers—where he makes rather detailed comparisons between the Begriffsschrift and Boole's "formula language" (Frege 1881/1979, 1882a/1979, 1882b/1972, 1883/1972). The background to this comprehensive attempt at demonstrating the superiority of the Begriffsschrift was a highly critical review of *Be griffsschrift* by Ernst Schröder, the foremost German proponent of Boolean algebraic logic (Schröder 1880/1972). Towards the end of the posthumously published "Boole's logical calculus and the Begriffsschrift", generally considered to be the most significant of these pieces, Frege summarises his findings with a list of six points. The first of these is that "My Begriffsschrift has a more far reaching aim than Boolean logic, in that it strives to make it possible to present a content when combined with arithmetical and geometrical signs". In order to realise these ambitions, the Begriffsschrift must be applicable, without any ambiguity, to the various sciences, and to mathematics in particular (we shall return to this point, but for another purpose). Frege demonstrates that the Boolean formula language, for several reasons, falls short of this ideal.[5] This is the sixth and final one:

> [My Begriffsschrift] can be used to solve the sort of problems Boole tackles, and even do so with fewer preliminary rules for computation. This is the point to which I attach least importance, since such problems will seldom, if ever, occur in science. (Frege 1881/1979, 47)

Here Frege informs the reader that in opposition to the Boolean calculus, the Begriffsschrift is developed, not with the purpose of solving pre-made puzzles, but

---

[4] Cf. Frege (1879a/1972, Preface, 1879b/1972), where the Begriffsschrift is used to express arithmetical and geometrical relations.

[5] This is made evident throughout the mentioned series of papers.

to play a definitive role in the sciences by contributing to the proofs of the validity of scientific judgements. The ideal justification of a scientific judgement is to derive it as a theorem of an axiomatic system (Frege 1879a/1972, 105–106).

Frege maintains that the Begriffsschrift is adequate to express the content of all of science, and that its inference rules have general applicability. This passage is illustrative:

> A few new signs suffice to present a wide variety of mathematical relations which it has hitherto only been possible to express in words [...]. But the usefulness of such formulae only fully emerges when they are used in working out inferences, and we can only fully appreciate their value in this regard with practice [...]. Nevertheless, the following example I have chosen may tempt people to experiment with the Begriffsschrift. It is of little significance which topic I choose, since the inference is always of the same sort and is always governed by the same few laws, whether one is working in the elementary or the advanced regions of the science. (Frege 1881/1979, 27)

Some 20 years later, in his critical treatment of Eduard Heine's and Johannes Thomae's formalism, Frege argues that "it is applicability alone which elevates arithmetic from a game to the rank of science" (Frege 1903a/1952, § 91).[6] A demonstration that arithmetic is logic is inconsequential and void of scientific value, unless it captures the applicability of arithmetic. It is clear, that from beginning to end, Frege held universal applicability to be a constitutive feature of logic and arithmetic. In fact, he thought of this as a strong argument in favour of the view that arithmetic is logic (Frege 1884/1968, § 13). Furthermore, for logic and arithmetic to have this property of universal applicability, they must themselves be full-fledged sciences.[7]

I have cited Frege's remark that "My Begriffsschrift has a more far reaching aim than Boolean logic, in that it strives to make it possible to present a content when combined with arithmetical and geometrical signs". One might think that this means that the Begriffsschrift has content only when applied to a science. The following passage has been taken to support such a reading:

> But [to express content] is exactly my intention. I wish to blend together the few symbols which I introduce and the symbols already available in mathematics, to form a single formula language. In it, the existing symbols [of mathematics] corresponds to the word-stems of [ordinary] language; while the symbols I add to them are comparable to the suffixes and the [deductive] formwords [*Formwörter*] that logically interrelate the content embedded in the stems. (Frege 1883/1972, 93)

---

[6] For a detailed treatment of Frege's objections towards formalism, cf. Heck (2010, 358–365), Costreie (2013) and Sören Stenlund's article (Chap. 5) in this volume.

[7] Cf. also "I now turn to the second of the two views that may be called formal theories [...]. This view has it that the signs of the numbers 1/2, 1/3, of the number π, etc. are empty signs. This cannot very well be extended to cover whole numbers, since in arithmetic, we cannot do without the content of the signs 1, 2, etc., and because otherwise no equation would have a sense which we could state—in which case we should have neither truths, nor a science, of arithmetic. It is curious, that it is precisely the lack of its consequential application, that has made the continuous existence of this opinion possible" (Frege 1885/1984, 114).

Michael Dummett illuminates the two facets nicely: "It is when he is criticising empiricism that Frege insists on the gulf between the senses of mathematical propositions and their applications; it is when he criticises formalism that he stresses that applicability is essential to mathematics" (Dummett 1991a, 60).

3 Frege's Unquestioned Starting Point: Logic as Science

In the second sentence, Frege presents the linguistic comparison that guides the alternative interpretation. The "But" at the beginning refers back to the fact that "the Boolean formula language as a whole [...] is a clothing of abstract logic in the dress of algebraic symbols. It is not suited for the rendering of a content" (Frege 1883/1972, 93).[8] The passage, together with some similar ones, occurs in the series of papers that Frege wrote in response to Schröder's review. The idea behind the reading is that Frege's understanding of logic switched from a formal Kantian approach, towards a substantial one, around the time of the publication of *Grundlagen*. However, the assumed emptiness of the "pure" Begriffsschrift is not at all Frege's point behind these passages. When applied to mathematics—arithmetic as well as geometry—the task of the Begriffsschrift is to capture logical interconnections among already appreciated scientific truths, or, in the rare case, to derive hitherto unknown scientific truths. Frege's point is to direct the attention to the superior deductive strength of the Begriffsschrift compared to the Boolean abstract logic. The flaw of Schröder's unqualified comparison of the two formula languages is evident from this remark, written several years after Frege gave up logicism: "*Mathematics has closer ties with logic than does any other discipline*; for almost the entire activity of the mathematician consists in drawing inferences" and "*inferring and defining are subject to logical laws*" (Frege 1914/1979, 203; italics original). Boole's formula language is incapable of capturing this close tie between mathematics and logic. Frege conveys this insight by the use of his apt linguistic comparison.

Let me substantiate this reading, by discussing the presentation of the Begriffsschrift that Frege provides, immediately after the passage under consideration. He begins with the content-stroke. It means "that the content which follows is unified, so that the other symbols can be related to it [as a whole]". By writing "$-4+2 = 7$" one simply makes it clear that "$4+2 = 7$" expresses a judgeable content. If one wishes "to assert a content as correct" one must add the judgment-stroke at the left end of the content-stroke. In order to make a correct judgment about $4+2 = 7$, one must negate it by way of the negation-sign, and then add the judgement-stroke, like this:

$\vdash_\top 4+2 = 7$.

After this clarification, Frege moves to a case involving the conditional-stroke. As always, he first presents his stipulation of it (discussed in some detail below), and then provides an example:

$\vdash (-x+2 = 4) \rightarrow (-x^2 = 4)$.[9]

---

[8] For a corresponding passage, cf. Frege (1881/1979, 13). Øystein Linnebo takes the first passage as evidence for the claim that Frege at the time subscribed to a Kantian formal understanding of logic (Linnebo 2003, 2013). In my view, he does not take into consideration the wider context, namely, the attempt on Frege's part to uncover the difference in ambition between his Begriffsschrift or formula-language and that of Boole.

[9] The expression contains three content-strokes; one connected to the judgement-stroke, and one preceding each of the arithmetical equations.

This judgement "denies the case that x^2 is not equal to 4 while nevertheless $x+2$ is equal to 4. We can translate it: if $x+2 = 4$, then $x^2 = 4$" (Frege 1883/1972, 94–95). This explication makes perfect sense of the linguistic comparison. The presentation of the judgement makes clear the logical relation between the antecedent and the consequent, both being arithmetic equations. When applied to arithmetic, or geometry for that matter, the Begriffsschrift makes evident the logical interrelationships between arithmetic or geometric judgeable contents. This is precisely because the logical signs, the judgement-stroke, the content-stroke, the negation-sign and the conditional-stroke, are sharply separated from each other and from the arithmetical signs. As underscored by the critical use of the comparison, this feature of the Begriffsschrift stands out in sharp contrast to the Boolean algebra. In the latter, algebraic and arithmetical signs are mingled together, and the very same signs sometimes have a logical sense, and sometimes have an arithmetical sense.

At another occasion, the early Frege formulates the very same demand for clarity:

> Anyone demanding the closest possible agreement between the relations of the signs and the relations of the things themselves will always feel it to be back to front when logic, whose concern is correct thinking and which is also the foundation of arithmetic, borrows its signs from arithmetic. To such a person it will seem more appropriate to develop for logic its own signs, derived from the nature of logic itself; we can then go on to use them throughout the other sciences [*anderen Wissenschaften*] wherever it is a question of preserving the formal validity of a chain of inference. (Frege 1881/1979, 12)

Viewed in this light, as simultaneously an expression of ambitions and a criticism of Boolean algebra, it is clear that the comparison does not address the question as to whether the Begriffsschrift itself has content. To conclude, a reading to the effect, that when he was working out the Begriffsschrift, Frege took it to be a formal notation, has little textual bearing. However, an overwhelming amount of textual evidence supports the ascription to Frege of a continuous view on these matters. Note, by the way, that Frege in the passage just cited, uses the phrase "the other sciences". This strongly suggests that logic is included among the sciences.

Frege has two tasks in mind when he develops the Begriffsschrift. In addition to presenting a notation or language that has universal applicability, he aims at formulating the most general systematic science. This science is an axiomatic system. Its core or basis is a minimal and, on all counts, a complete set of axioms.[10] I shall discuss the status of the axioms below, but let me first spell out some general features of logic as science. In a characteristic manner, Frege states that "The goal of scientific endeavour is *truth*. Inwardly to *recognize something as true is to make a judgement*, and to give expression to this judgement is to make an assertion" (Frege 1880?/1979, 2; italics original). A systematic scientific theory—any systematic scientific theory—consists of asserted judgements. After he introduced the sense-meaning distinction, Frege repeatedly drew the distinction between scientific and non-scientific activity by stating that the latter might stick to the thought, while the former always involves an additional struggle for meaning (Frege 1897b, 129–130,

---

[10] Frege lacked a method for proving completeness, cf. Frege (1879a/1972, § 13, 1897a/1984, 235).

1906b/1979, 191). In "On Sense and Meaning", we twice encounter the metaphor that "judgements can be regarded as advances from a thought to a truth value" (Frege 1892a/1984, 164–65 and 177).[11] In light of this general and unrestricted description, we might say with Joan Weiner that "There is, for Frege, only one notion of truth. It is what we obtain in systematic science" (Weiner 2010, 49).[12] That is to say, we do not have truth by convention or truth in virtue of meaning, on the one hand, and material truth, on the other: To acknowledge a thought to be true is to make a judgement.

Frege's unwavering position is evident from his diagnosis of his well-known "encounter" with the psychological logician in the preface to *Grundgesetze*:

> Surveying the whole question, it seems to me that the source of the dispute lies in a difference in our conceptions of what is true. For me, what is true is something objective and independent of the judging subject; for psychological logicians it is not [...].
>
> We can generalize this still further: for me there is a domain of what is objective, which is distinct from that of what is actual, whereas the psychological logicians, without ado take what is not actual to be subjective. And yet it is quite impossible to understand why something that has a status independent of the judging subject has to be actual, i.e., has to be capable of acting directly or indirectly on the senses. (Frege 1893/1967, xvii–xviii)

In Frege's mind, to make a judgement is to judge a thought to be true and that which is true is true independently of the judging subject. If we relate to the different understandings of the Fregean logicism mentioned at the outset of this article, we could say that this presupposition of objectivity—covering both abstract objects, such as the individual numbers, and thoughts and their constituents—captures Frege's metaphysical presupposition. It plays a fundamental role in his numerous attacks on psychologism and formalism, and he maintains that without objectivity, we would have no science (Frege 1880?/1979, 6, 1892b/1984, 185(n), 1897b/1979, 133). The metaphysical readings referred to in the introduction, ascribe a far richer metaphysics than this presupposition of objectivity to Frege.[13]

---

[11] Cf. Frege (1880?/1979, 7, 1897b/1979, 139, 1906b/1979, 185, 1906c/1979, 198, 1919/1979, 253, 1925/1979, 267). To Frege, after introducing the sense-meaning distinction, a sentence is a name that expresses a thought and means an object, either the True or the False. I shall refrain from raising the complex issue as to whether Frege's position is adequately captured by the label "realism". For two competing accounts, cf. Carl (1994, 137–160) and Dummett (1981, 428–473).

The reson I refer to "Logic" by "1880?" is that the editors of Frege, *Posthumous Writings* are uncertain about when it was drafted; they suggest sometimes between 1879 and 1891.

[12] In the normative phrasing of Gilead Bar-Elli, Frege, in common with Quine, subscribes to the "homogeneity of truth" principle: "Whatever your metaphysical view of truth is, it should not recognize radically different kinds or notions of truth but should be homogenous across the board" (Bar-Elli 2010, 180). Thomas Ricketts uses the phrase "monolithic view of truth" (Ricketts 1996, 126–127).

[13] Jeshion ascribes to Frege "a foundational hierarchy" of propositions (Joshin 2001, 939), the view that there are simples that "constitute the essence of a discipline" and an idea about "the natural order of truth" (Jeshion 2001, 947). Steven Shapiro, in an account deeply influenced by Jeshion, label their common reading a "metaphysical-cum-epistemic account" (Shapiro 2009, 184), and he ascribes to Frege "a large dose of pre-established harmony" (Shapiro 2009, 185). To ascribe to Frege such a "thick" rationalistic and pre-Kantian metaphysics—unprovable and speculative as it is—goes against the view on science and justification formulated here.

We have looked at two characteristics of the sciences: The first is that they are interconnected in a number of ways, and that logic and arithmetic are applicable to all sciences. The second, is that a scientific activity consists in grasping thoughts, in making judgements, and in assertions. We should also add a third characteristic: Each science presupposes elucidations and pre-scientific clarifications of basic terms. This is evident from Frege's remark that "Logic, in common with every science, has its technical terms, words some of which are also used outside the sciences, though not in quite the same sense" (Frege 1880?/1979, 5).[14] Frege substantiates this condensed remark in his most concise account of the basic ingredients of science, the unpublished "Logic in Mathematics" from 1914:

> Definitions proper must be distinguished from *elucidations* [*Erläuterungen*]. In the first stages of any discipline [*Wissenschaft*] we cannot avoid the use of ordinary words. But these words are, for the most part, not really appropriate for scientific purposes, because they are not precise enough and fluctuate in their use. Science needs technical terms that have precise and fixed meanings, and in order to come to an understanding about these meanings and exclude possible misunderstandings, we elucidate their use. Of course in so doing we have again to use ordinary words, and these may display defects similar to those which the elucidations are intended to remove. So it seems that we shall then have to do the same thing over again, providing new elucidations. Theoretically one will never really achieve one's goal in this way. In practice, however, we do manage to come to an understanding about the meanings of words. Of course we have to be able to count on a meeting of minds, on others guessing what we have in mind. But all this precedes the construction of a system and does not belong with a system. (Frege 1914/1979, 207)[15]

To avoid future misunderstandings, I need to distinguish between two kinds of Fregean stipulations. The first is the one discussed in this passage, that is, elucidations of the use of a science's basic technical terms. Our discussion of the content-stroke, the judgement-stroke, and the conditional-stroke belong to this category. The other kind of stipulation, mentioned at the beginning of the passage, is definition proper [*Eigentliche Definition*]. These introduce new signs into an already established science. Their sole role is to facilitate proofs, and, in principle, they are entirely dispensable (Frege 1914/1979, 224). A definition is neither true nor false, but might be transformed into an analytic judgement.

---

[14] Cf. Weiner (1990, 230) and Ricketts (2010, 191–192). As Frege says in "On the Foundations of Geometry: Second Series": "Just as the concept point belongs to geometry, so logic, too, has its own concepts and relations; and it is only in virtue of this that it can have a content [...] to logic for instance, there belongs the following: negation, identity, subsumption, subordination of concepts" (Frege 1906a/1984, 338). John MacFarlane (2002, 33) invokes this passage in his criticism of Ricketts's ascription to Frege of the view that, "in contrast to the laws of special sciences like geometry or physics, the laws of logic do not mention this or that thing. Nor do they mention properties whose investigation pertains to a particular discipline" (Ricketts 1985, 4–5). MacFarlane, however, misses the point, as no serious scholar, certainly not Ricketts, would ever dream of denying that there are logical constants; see for instance Ricketts (1996, 123).

[15] In their respective articles in the present volume, Gisela S. Bengtsson (Chap. 7) and Gottfried Gabriel (Chap. 2) prefer other renderings of "verständnisvollen Entgegenkommen" than "meeting of minds". Bengtsson prefers "willingness to understand", and Gabriel, "accommodating understanding". I have changed the translation of "Erläuterungen" to "elucidate" and "elucidations" rather than keeping "illustrative examples" and "new meanings".

In the next part, we shall look at selected aspects of Frege's two systems of the most general science, that of *Begriffsschrift* and *Grundgesetze*, respectively.

3.3 Logic as Systematic Science and Philosophy

Frege observes that:

> To make a judgement because we are cognisant of other truths as providing a justification for it is known as *inferring*. There are laws governing this kind of justification, and to set up these laws of valid inferences is the goal of logic (Frege 1880?/1979, 3).

We have discussed inferring with respect to judgements in arithmetic, geometry, and other sciences, and we have noted that the project of *Begriffsschrift* is to present the most general systematic science. What about the status of the axioms of this system, that is, the non-inferred judgements? Our task is to flesh out the following passage:

> Now the grounds which justify the recognition of a truth often reside in other truths which have already been recognized. But if there are any truths recognized by us at all, this cannot be the only form that justification takes. There must be judgements whose justification rests on something else, if they stand in need of justification at all.
>
> And this is where epistemology [*Erkenntnistheorie*] comes in. Logic is concerned only with those grounds of judgement which are truths. (Frege 1880?/1979, 3)

When the justification of a theorem is traced all the way back to the primitive truths that ground its derivation, we have reached the limits of logical justification.[16] This condition holds for all science, including logic.[17] Frege makes it abundantly clear that logical axioms or primitive truths cannot be justified by any of the restricted sciences "that themselves require fundamental logical principles [*logischen Grundsätze*]" (Frege 1893/1967, xix).[18]

---

[16] Frege is fully aware that there are always alternative axiomatic systems, so that a judgement could be an axiom in one system and a theorem in another (Frege 1879a/1972, § 7; 1914/1979, 205). I cannot see that this poses a problem for the present reading, as Frege does not accept a holistic or pragmatic strategy for justifying axioms; cf. Shapiro (2009).

[17] For the term "primitive truth", cf. Frege (1914/1979, 204).

[18] The full passage is cited in Ricketts (1996, 126); Gisela Bengtsson drew it to my attention. The very same allusion (although concerning arithmetic and not logic) is made also in another context, namely in the criticism of formalism:

> A clear-cut separation of the domains of the sciences may be a good thing, provided no domain remains for which no one is responsible. We know that the same quantitative ratio (the same number) may arise with lengths, time interval, masses, moments of inertia, etc.; and for this reason it is likely that the problem of the usefulness of arithmetic is to be solved—in part, at least—independently of those sciences to which it is to be applied. Therefore, it is reasonable to ask the arithmetician to undertake the task, so far as he can accomplish it without encroaching on the domains of the other special sciences. To this end it is necessary, above all things, that the arithmetician attach a sense to his formulas; and this will then be so general that, with the aid of geometrical axioms and physical and astronomical observations and hypotheses, manifold applications can be made to the sciences. (Frege 1903/1952, § 92)

Let us begin with Frege's response to this challenge about the status of the axioms in the axiomatic system of *Begriffsschrift*. Frege specifies a notion of content, that of "conceptual content", in *Begriffsschrift* § § 2–3. It concerns solely the influence on possible inferences; all other aspects of meaning, significance, or content, are left out of the picture as irrelevant to logic. A content "that can become a judgement" is, as we have seen, called a "judgeable content". Then, in *Begriffsschrift* § 5, Frege introduces the conditional-stroke, together with this stipulation of conceptual content:

If A and B stand for judgeable contents, there are the following four possibilitites:

1. A is affirmed and B is affirmed
2. A is affirmed and B is denied
3. A is denied and B is affirmed
4. A is denied and B is denied.

Now, ⊢(—B→—A) stand for the judgement that the third of these possibilities does not occur, but one of the other three does.

On this basis, Frege presents his first axiom, ⊢(—a→(—b→—a)), which is Formula (1) of the *Begriffsschrift* (Frege 1879a/1972, § 14).[19] In order to situate Frege's discussion of Formula (1) correctly, I will utilise a distinction that Frege introduces in *Grundgesetze*, namely that between *Zerlegung* and *Aufbau*. A *Zerlegung* is an explanation, or an elucidation in quasi-colloquial language, of a given axiom or theorem. It involves the use of words. In contrast, an *Aufbau* consists of symbols only; it is the symbolic presentation of an axiom or the derivation of a theorem. Although Frege had not yet labelled this distinction in 1879, a glance at the proofs in *Begriffsschrift*, Part II: "Representation and Derivation of some Judgements of Pure Thought", shows that it was in play back then. We have every reason to treat the following two remarks as *Zerlegungen*: "[Formula 1] says: 'The case in which a is denied, b is affirmed and a is affirmed is excluded'" and "If a proposition a holds, it holds also in case an arbitrary proposition b holds". On this basis, Frege remarks that his first axiom "is obvious since a cannot be denied and affirmed at the same time" (Frege 1879a/1972, § 14). We might summarise these *Zerlegungen* by a comprehensive one: "In case you do not yet grasp the judgeable content expressed by Formula (1), it might help to note that it cannot consistently be denied, thus it must be affirmed". If this is correct, then it is clear that by his *Zerlegungen*, Frege spells out the content of Formula (1) for the reader not yet fully familiar with the notion of a judgeable content, or the stipulated conceptual content of the conditional-stroke, or both. In light of this, I see no reason to reflect further upon Frege's notion of "obvious". It simply forms part of a comprehensive *Zerlegung* and carries no substantial or theoretical commitment.[20] The presentation of the rest of the axioms of

---

[19] The alternation of capital and small letters between the explication and the axiom is because in the former case a relation is introduced, while in the latter case we have an assertion.

[20] Among the purposes of the present reading of Frege is to avoid ascribing to him highly sophisticated notions of "self-evidence", as in Jeshion (2001) and Shapiro (2009). Such ascriptions carry with them a full-fledged metaphysical interpretation of the Fregean logicism; cf. note 13.

the *Begriffsschrift* follows the same pattern. Frege thought, then, that the axioms of the *Begriffsschrift* need no, and cannot be given any, justification. The *Zerlegungen* are explications, *Erläuterungen*, and not justifications.

It might be tempting to take Frege to claim that to grasp the judgeable content of the axioms—with or without the help of the *Zerlegungen*—is to realise that they are true. Since to realise that something is true is to make a judgement, this would mean that to grasp the content of an axiom is to judge it to be true. Stated the other way around: Someone who does not judge an axiom to be true, has not fully or adequately grasped its content. But this reading is too strong, as it provides an explanation of a particular class of judgements. Frege, however, writes that "Judgements can be regarded as advances from a thought to a truth-value. Naturally, this cannot be a definition. Judgement is something quite peculiar and incomparable" (Frege 1892a/1984, 164–165). We cannot give a reductive account of a judgement, whether in light of some notion of epistemic or metaphysical necessity, or by way of causality.[21] Let us once again turn to Frege's encounter with the psychological logician, this time in order to enlighten the claim that we have reached bedrock. Let us begin with this passage:

> But what if beings were even found whose laws of thought flatly contradicted ours […]? The psychological logician could only acknowledge the fact and say simply: those laws hold for them, these laws hold for us. I should say: we have here a hitherto unknown type of madness. (Frege 1893/1967, xvi)

In opposition to the psychological logician, Frege takes our laws of thought to prescribe how we ought to think. He takes side and maintains, that those who follow our laws are right, the others are wrong. However, this is the limit of logical justification:

> The question why and with what right we acknowledge a law of logic to be true, logic can answer only by reducing it to another law of logic. Where that is not possible, logic can give no answer. If we step aside from logic, we may say: we are compelled to make judgements by our own nature and by external circumstances; and if we do so, we cannot reject this law—of Identity, for example; we must acknowledge it unless we wish to reduce our thought to confusions and finally renounce all judgement whatsoever. I shall neither dispute nor support this view; I shall merely remark that what we have here is not a logical consequence. What is given is not a reason for something's being true, but for our taking it to be true. Not only that: this impossibility of our rejecting the law in question hinders us not at all in supposing beings who do reject it; where it hinders us is in supposing that these beings are right in so doing, it hinders us in having doubt whether we or they are right. At least this is true of myself. (Frege 1893/1967, xvii)

Although other sciences than logic, evolutionary psychology, or biology, say, might provide well-based reasons for our holding a logical law to be true, they cannot state why they are true. Since such reasons cannot be provided by logic either, they cannot be provided by science at all. We might, then, imagine that there are beings that grasp the content of Formula 1 and make the judgement that its negation is true.

---

[21] On the confusions involved in the latter idea, cf. Frege (1880?/1979, 2–3).

This possibility demonstrates that there is a principled advance from grasping a thought to making a judgement.[22]

Frege says that if we reject a logical law, we "reduce our thought to confusion". In a passage from *Grundlagen*, aiming at motivating logicism, he argues that to try to deny one "of the fundamental propositions of the science of numbers" leads to complete confusion and "Even to think at all seems no longer possible" (Frege 1884/1968 § 14).[23] Such remarks mark a clear difference between logic and arithmetic on the one hand, and the rest of the sciences, on the other. For the purpose of a conceptual investigation one could, for instance, negate one of the Euclidian axioms. On this basis, Frege asks the rhetoric question: "Should not the laws of numbers, then, be connected very intimately with the laws of thought?" (Frege 1884/1968, § 14). Frege's claim is strong: to deny a logical law leads to complete confusion, and it seems that we would then not even be thinking. However, as argued, this claim is not a scientific one; but a philosophical or epistemological one. Formulated differently: Since logic is the most general science, and cannot justify its own basic judgements, these judgements cannot be justified by science.[24]

What about the axioms of the more complex axiom-set of *Grundgesetze*; are they self-evident as well? Since the principal new axioms, axioms V and VI, have the identity-sign as the main connective, some elaboration is needed before we are in the position to respond adequately to this question. Our focus is on the fifth one:

(V) $\vdash(\acute{\epsilon}f(\epsilon) = \acute{\alpha}f(\alpha)) = ((x)(f(x) = g(x)))$.
(V) Expresses an identity between the pairs

(a) The value-range of f = the value-range of g
(b) For every x, $f(x) = g(x)$.

For our purposes, it is enough to note that the notion of a value-range is a generalisation of that of an extension (Frege 1893/1967 § 9).

Axiom V is analytic, but it is not obtained from a definition.[25] It is not possible to derive one of the sides from the other, and both are needed. The role of Axiom V is to enable us to form logical proper names in the Begriffsschrift by way of value-ranges.[26] Clearly, in his attempt at comprehending the epistemological nature of Axiom V, Frege must have reflected deeply on the content and significance of

---

[22] This passage stands in debt to comments by Joan Weiner.

[23] Here, as in the passage from *Grundgesetze* looked at above, the German "Verwirrung" is translated as "confusion".

[24] M. Dummett and R. Heck relate the issue about the justification of the logical law to the familiar Cartesian Circle, according to which an attempt at justifying a logical axiom must itself presuppose logic. Both try to escape, or at least minimise the damage of the circle by way of semantic considerations (Dummett 1991b, 200–215; Heck 2010, 342–358). In opposition to this, Frege thinks that his basic laws cannot, and need not be justified, but might be elucidated. Accordingly, we should not ascribe to Frege the task that occupies these semanticists.

[25] There is an illuminating discussion about the non-definitional, non-stipulative status of Axiom V in Frege (1903a/1952, § 146).

[26] Cf. Frege (1893/1967, § 9; 1903a/1952, § 147).

3 Frege's Unquestioned Starting Point: Logic as Science

identity-judgements. In the introduction to *Grundgesetze*, he notes that sometimes after *Begriffsschrift* his understanding of (logical) identity switched to the ordinary mathematical one (Frege 1893/1967, ix).[27] Frege faces a task of considerable complexity: A satisfying analysis of the identity-relation must take account of the fact that identity-judgements might improve knowledge *and* that the linguistic form of an identity-judgement is not decisive for the question as to whether it is analytic or synthetic; Axiom V has the form "$a=b$".

At the beginning of "On Sense and Meaning", Frege presents a distinction that any adequate analysis of identity must capture:

> a=a and a=b are obviously sentences of differing epistemic value [*Erkenntniswert*], a=a holds *a priori* and, according to Kant, is to be labelled analytic, while sentences of the form a=b often contain very valuable extensions of our knowledge [*sehr wertvolle Erweiterungen unserer Erkenntnis*] and cannot always be established [*nicht immer zu begründen sind*] *a priori*. (Frege 1892a/1984, 157)[28]

In arithmetic and geometry, solving equations is a principal task. From a broadly Kantian perspective, the identity-judgements thus made are *a priori*. (As is well known, Frege does not take empiricism in the philosophy of mathematics seriously.) However, Frege exemplifies the distinction he is after, by way of an astronomical example, and not a mathematical one. This is a wise choice, since he then avoids a number of irrelevant philosophical controversies. It is evident that the judgement that the Morning Star is the Morning Star has little importance for our knowledge and that the judgement that the Moring Star is the Evening Star improves knowledge—it even led to the introduction of a new name, "Aphrodite's Star". We need an explanation of what it is about identity-judgements that makes this huge difference with respect to knowledge, some are trivial and some move science forward. The introductory passage and the astronomical example together make it clear that the challenge is to give an adequate account of the notion of epistemic value. The Kantian notions of analytic and *a priori* play a secondary role.

Towards the very end of "On Sense and Meaning", Frege summarises his solution, based on the two-levelled notion of content introduced by the title:

> If we found '$a=a$' and '$a=b$' to have different epistemic values, the explanation is that for the epistemic value [*für den Erkenntniswert*], the sense of the sentence, viz. the thought expressed by it, is no less relevant than its meaning, i.e. its truth-value. If now $a=b$, then indeed what is meant by 'b' is the same as what is meant by 'a', and hence the truth-value of '$a=b$' is the same as that of '$a=a$'. In spite of this, the sense of 'b' may differ from that of 'a', and thereby the thought expressed in '$a=b$' differs from that of '$a=a$'. In that case the

---

[27] In his *Begriffsschrift* account of identity, Frege maintained that "Identity of content differs from conditionality and negation by relating to names, not to contents", and that "with the introduction of a symbol for identity of content, a bifurcation is necessarily introduced into the meaning of every symbol" (Frege 1879a/1972, § 8). By the new account, this bifurcation of meaning is avoided.

[28] I have modified Max Black's translation by translating "Erkenntniswert" as "epistemic value" and not "cognitive value"—this translation connects to the numerous places where Frege uses "Erkenntnis", "Erkenntnistheorie" and constructions such as "kennst man"—and by rendering "Sätze" as "sentences" rather than "statements".

two sentences do not have the same epistemic value. If we understand by 'judgement' the advance from the thought to its truth-value [...], we can also say that the judgements are different. (Frege 1892a/1984, 177)[29]

Frege now thinks that the problem with his original account of identity (Frege 1879a/1972, § 8) is that it did not enable us to "express proper knowledge" [*eigentliche Erkenntnis darin ausdrücken*] (Frege 1892a/1984, 157).[30] Without sense, "we should have no knowledge at all"; all sentences would have the same epistemic value as "*a=a*" (Frege 1904/1980, 164–65). And without the distinction between sense and meaning, as stated in a letter to Russell, and made evident by the term "*Erkenntniswert*", we would be in no position to make sense of the fact that we sometimes need both investigations and reflections before we make the judgement that a certain thought is true (Frege 1904/1980, 164–65).

In his concluding passage, Frege does not invoke the Kantian notions. To repeat, this is because his solution to the identity-problem is independent of the status of identity-judgements as *a priori* or *a posteriori*. His claim is twofold: First, if "*a*" and "*b*" have or express the same sense, then "*a=a*" and "*a=b*" express the same thought, and if "*a*" and "*b*" have different senses, then "*a=a*" and "*a=b*" express different thoughts. In other words, "*a=a*" and "*a=b*" express different thoughts, if and only if, "*a*" and "*b*" have different senses. For our subsequent discussion, it is crucial to note that to say that a judgement is of the form "*a=b*" is to make a linguistic classification, it does not address the sameness or difference of the sense or senses expressed by "a" and "b".[31] Second, since Frege's discussion centres around epistemic value and the significance of judgements for knowledge and the progress of science, it is presupposed in his analysis that in fact *a=b*. Of course, there are false thoughts of the form "a=b", but these do not enter the picture.[32]

In the final sentence of the concluding passage, Frege states, that given his understanding of judgement, if two identities have different epistemic values, the corresponding judgements differ. This might encourage the reading that this difference is captured by the Kantian distinctions, and that there is a systematic correspondence between differences in epistemic values and the grouping of judgements into analytic, synthetic, *a priori* or *a posteriori*. Wolfgang Carl forcefully argues for this interpretation in a work that shares the general outlook of the present article and volume. "My aim", says Carl, is to present a reading that will "illustrate and support

---

[29] Cf. also Frege (1903a, § 138).

[30] I here sidestep the rather complex issue as to whether Frege's reconstruction of his former account of identity is adequate.

[31] The German original of, "In spite of this, the sense of '*b*' may differ from that of '*a*', and thereby the thought expressed in '*a* = *b*' differs from that of '*a* = *a*'" is "Trotzdem kann der Sinn von "*b*" von dem Sinn von "*a*" verschieden sein, und mithin auch der in "*a* = *b*" ausgedrückte Gedanke verschieden von dem [in] "*a* = *a*" ausgedrückten sein". Cf. also the first sentence of the concluded passage of "On Sense and Meaning", cited above.

[32] Within the context of the philosophy of propositional attitudes and other (purported) intensional contexts, this crucial presupposition for epistemic value is often overlooked. Nathan Salmon, for instance, translates *Erkenntniswert* by the highly loaded, semantic, and misleading "cognitive information content" (Salmon 1986, 47).

3 Frege's Unquestioned Starting Point: Logic as Science

the view that Frege's theory of sense [*Sinn*] and reference [*Bedeutung*] belongs to the epistemological tradition of modern philosophy that stems from Kant" (Carl 1994, vii). Before we continue with our main subject at present, the epistemic status of Axiom V, I would like to mention that I think Carl's understanding of the notion of epistemic value, as formulated in this passage, is misguided:

> The origin of the idea of the cognitive value [*Erkenntniswert*] of a sentence can be traced back to *The Foundations of Arithmetic*, where Frege alludes to the distinction between *a priori* and *a posteriori* judgements, taken over from Kant, does not refer to "the content of the judgement but to the justification for making the judgement" and to "the ultimate ground upon which the justification for holding it true rests" (§ 3). In "On Sense and Reference" he claims that an *a priori* sentence differs from an *a posteriori* one by having a different "cognitive value" […]. There is no reason to suppose that he ever changed his mind about the interpretation of the Kantian distinction […].
>
> If the distinction between a priori and a posteriori judgements concerns the difference in their mode of justification, and if this difference leads to a difference in cognitive value of the sentences used to express these judgements, then a difference in justification is a sufficient condition for a difference in cognitive value. (Carl 1994, 147)

According to my reading, "epistemic value" simply means "value for our knowledge". Some judgements have great value for our knowledge; others have little or are void of value. Frege uses the term "epistemic value" to set the stage for the introduction of the distinction between sense and meaning. Thus, *pace* Carl, the difference between a judgement that has little epistemic value and a judgment that has great epistemic value coincides neither with the distinction between the *a priori* and the *a posteriori*, nor with the one between the analytic and the synthetic.[33] This open-ended and non-committed reading of the notion of epistemic value is vindicated by a short discussion in "On the Foundations of Geometry: First Series" (Frege 1903b/1984). Frege begins his polemic article, directed at the prevalent influence of Hilbert in the studies of the foundations of mathematics, by presenting his own view and locating himself as an heir of a dignified tradition. "Traditionally", he observes, "what is called an axiom is a thought whose truth is certain without, however, being provable by a chain of logical inferences" (Frege 1903b/1984, 273).[34] Then he observes that "Once a word has been given a meaning by means of a definition, we may form a self-evident proposition [*einen selbstverständlichen Satz machen*] from this definition, which may then be used in constructing proofs in the same way in which we use principles" (Frege 1903b/1984, 274).[35] Immediately after these clarifications, Frege invokes the notion of epistemic value:

> [E]ven if what a definition has stipulated is subsequently expressed as an assertion, still its epistemic value is no greater than [*nicht größer als*] that of an example of the law of identity $a=a$. By defining, no knowledge is engendered; and thus one can only say that definitions

---

[33] Cf. Frege (1903b/1984, 274), where Frege talks about one epistemic value as "no greater" than another; see also the quote below.

[34] For a simple and clear presentation of the different account of axioms by Frege and Hilbert, cf. Eder (2015). (I do not agree to all of the author's claims about Frege and independence proofs, but that is more or less independent of the exposition of the basic disagreement at stake.)

[35] Cf. the aspect about kinds of stipulations towards the end of Part II.

that have been altered into assertoric propositions formally play the role of principles but really are not principles at all. For although one could just possibly [*allenfalls*] call the law of identity itself an axiom, still one would hardly wish to accord the status of an axiom to every single instance, to every example, of the law. For this, after all, greater epistemic value [*einen größeren Erkenntniswert*] is required. (Frege 1903b/1984, 274)

Although this passage is condensed, and opens the field for rather complex issues concerning Frege's understanding of the role of stipulative definitions in axiomatic systems,[36] it is clear that he maintains that an axiom has greater epistemic value than an assertion obtained from a definition, or just an arbitrary instance of the law of identity. It is furthermore explicitly stated that the principle of identity has a greater epistemic value that any of its instances, but still not a very high one. As this principle belongs to pure logic, and accordingly, is analytic, it is clear that analytic judgements might have at least some epistemic value.

We have seen that if judgements differ in epistemic value, they are different judgements. This way of identifying judgements concerns the value of the judgement for our knowledge. The problem with Carl's reading is that the Kantian distinctions are absolute, while differences in epistemic value are gradual. A judgement is either analytic or not analytic. A judgement might have more or less epistemic value than another, and two analytic judgements, viz., the principle of identity and one of its instance, might differ in epistemic value. Clearly, then, classifications of judgements according to Kantian epistemological notions and in terms of epistemic value, respectively, are independent of each other.[37]

Now, back to our main topic: In the final sentence of the passage in which Frege formulates his puzzle, it is stated that "$a=b$" "cannot always be established [*begründen*] *a priori*". Since Axiom V is not inferred from prior judgements, it is not proved. This means that it must be "completely self-evident" [*ganz selbstverständlich*] (Frege 1903a/1952, § 60). In the relevant sense, this identity is not established *a priori*, as it is not established at all. Furthermore, general agreement and wide applicability cannot take care of this self-evidence, "in mathematics a mere moral conviction, supported by a mass of successful applications, is not good enough" (Frege 1884/1969, § 1). Axiom V is of the linguistic form "$a=b$". From what we have just said, it is clear that if the two sides express different senses, then some deductive reasoning is needed in order to establish [*begründen*] the judgement that $a=b$. However, this faces the claim that in order to be an axiom, Axiom V must be "completely self-evident". It seems to me that the conclusion is unavoidable: Despite being different expressions, or having different forms, "For every $x, f(x) = g(x)$", and "The value-range of f = the value-range of g", express the same sense. Accordingly, Frege thought that to grasp the sense of the two expressions flanking the identity sign of Axiom V is sufficient to make the judgement that it is true, and conversely, if one is uncertain about the truth-value of Axiom V, then one has not fully grasped the thought that it expresses.

---

[36] Cf. Floyd (1998).

[37] Towards the end of this part, I present my understanding of the role of the Kantian distinctions in Frege's philosophy.

3 Frege's Unquestioned Starting Point: Logic as Science

The ascription to Frege of the view that the self-evidence of the axioms *Grundgesetze* is made sense of, in terms of a set of stipulations, and the notions of sense and meaning gain support from his account of the Euclidian axiom of parallels:

> When a straight line intersects one of two parallel lines, does it always intersect the other? This question, strictly speaking, is one that each person can only answer for himself. I can only say: so long as I understand the words 'straight line', 'parallel' and 'intersect' as I do, I cannot but accept the parallels axiom. If someone else does not accept it, I can only assume that he understands these words differently. Their sense is indissolubly bound up with the axioms of parallels. (Frege 1914/1979, 247)[38]

Frege introduces Axiom V as a logical axiom in "Function and Concept" and the structural similarity to this passage is striking. Frege first elucidates his understanding of the words "function", "value", "generality", "extension", and "graph", and thereafter, on that basis, maintains that the two sides of Axiom V express the same sense (Frege 1891/1984, 143). Frege could have put it like this: "So long as I understand the words 'function', 'value', 'generality' and 'graph' as I do, I cannot but accept Axiom V". A substantial justification is neither needed, nor possible.[39] It does not follow from the present account, however, that if an identity is of the form "$a=b$", and "a" and "b" have different senses, then the corresponding judgement is synthetic. The reason is that such an identity-judgement might be derivable as a theorem in the system of *Begriffsschrift* or *Grundgesetze*, and in that case, it is analytic. We might make an identity-judgement and be uncertain, or even wrong, about its status as analytic or synthetic (*a priori*).[40]

In response to the question about the justification, or need for justification of the axioms of *Grundgesetze*, we have found that Frege has not changed his view. He always held the view that with respect to axioms, a grasp of content is sufficient for making the judgement that it is true; reasoning is not involved. However, the *Zerlegungen* differ considerably for two reasons. First, the invoked notion of content is, as we have seen, far more complex after the introduction of the sense-meaning distinction, and, second, due to the introduction of the sign for value-ranges and the definite description operator, the required stipulations are considerably more complex in *Grundgesetze*.[41] ("Stipulation" is here used in the sense of elucidation of meaning and not in the sense of definition proper.) To make my view clear, some further reflections on these stipulations are required. We have looked at the stipulation of the conceptual content of the conditional-stroke. In *Grundgesetze*, Frege provides a stipulation for the value-range:

> I use the words
> "the function $\Phi(\xi)$ has the same value-range as the function $\psi(\xi)$"

---

[38] Cf. also Frege (1903b/1984, 278, 1906a/1984, 333–334), taken together.

[39] Cf. Frege (1914/1979, 205). Frege's discussion about the status of Axiom V is always a bit too insisting; cf. Frege (1893, Introduction, 1903a/1952, § § 146–147). The line of interpretation presented here is spelt out in detail in Alnes (1999, 2013).

[40] Cf. the discussion about *Begriffsschrift* formula 133, below.

[41] Cf. Frege (1893/1967, § § 3, 9 and 11). A major consequence of these new complex requirements of the *Zerlegungen* is that they no longer carry the strong intuitive appeal of those of *Begriffsschrift*.

generally to denote the same as [*gleichbedeutend*] the words
"the functions Φ(ξ) and ψ(ξ) have always the same value for the same argument". (Frege 1893/1967, § 3)

A few sections later, Frege refers back to this as a stipulation [*Bestimmung*] (Frege 1893/1967 § 9). When he introduces Axiom V, Frege simply refers to these former stipulations (Frege 1893/1967, § 20). It is, however, crucial, as in the case of Axiom 1—Formula (1)—of *Begriffsschrift*, not to take this to mean that Frege ever thought that Axiom V is true by stipulation or by convention.[42] As I hope to have made clear, a grasp of the exact content of a stipulation might be a presupposition for making a judgement, and rather comprehensive *Zerlegungen* (including elucidations) might be needed in order to clarify this content. The ascription to Frege of the outlined founding of an axiom or a primitive truth gains considerable support from this remark from around 1923: "the truth of a logical law is immediately evident of itself, from the sense it expresses" (Frege 1926/1984, 405).

We are now deeply into philosophy. When we classify a scientific judgement as analytic, we rely on the notions of sense and meaning, a set of stipulations, and eventually a circumscribed set of inference rules. Furthermore, it is clear from our discussion of the first and final passages of "On Sense and Meaning" that the notions of sense and meaning belong to epistemology. From all this, it follows that the classification of judgements as analytic, synthetic, *a priori* or *a posteriori* does not form part of the scientific enterprise.

Frege paraphrases the final judgement of *Begriffsschrift*, formula 133, like this: "If the procedure *f* is many-one, and if *m* and *y* follow *x* in the *f*-sequence; then *y* belongs to the *f*-sequence beginning with *m*, or precedes *m* in the *f*-sequence" (Frege 1879a/1972, § 31).[43] The proof that this judgement is analytic is highly important to Frege.[44] As he certainly expected, the claim that it is analytic met with distrust. Nobody, of course, questioned the correctness of the judgement; but it was a common opinion that notions such as "successor" and "follows in a series" are based in intuition, and, accordingly, that the judgment is synthetic.[45] In one of the

---

[42] Cf. note 12 and the discussion leading up to it.

[43] In *Grundlagen*, Frege paraphrases the judgement thus: "If the relation of every member of a series to its successor is one-or many-one, and if *m* and *y* follow in that series after *x*, then either *y* comes in that series before *m*, or it coincides with *m*, or it follows after *m*" (Frege 1884/1968, § 91).

[44] Cf. Frege (1879a/1972, § 23; 1881/1979, 38).

[45] As said in one of the reviews of *Begriffsschrift*:

> I cannot completely agree with the comments which Frege makes about the relations which the fundamental concepts of mathematics have to each other. I cannot agree that the concept of ordering-in-a-sequence can be reduced to that of logical ordering, let alone the concept of number can be advanced by investigations into ordering-in-a-sequence. On the contrary, the concept of ordering-in-a-sequence is a secondary one, dependent upon the concept of time; while the concept of number is a primary mathematical one—indeed the simplest, most general concept of all (Michaëlis 1880/1972, 217).

C. Th. Michaëlis not only takes formula 133 to be synthetic and based in intuition, he further warns Frege against pursuing the project announced in the introduction to *Begriffsschrift* (cited in note 2 above).

3 Frege's Unquestioned Starting Point: Logic as Science 41

concluding passages of *Grundlagen*, pointing back to this particular proof, Frege could in triumph maintain that "From this proof it can be seen that propositions which extend our knowledge can have analytic judgements for their content" [*Aus diesem Beweise kann man ersehen, dass Sätze, welche unsere Kenntnisse erweitern, analytische Urtheile enthalten können*] (Frege 1884/1968, § 91). The message is clear: Kant's way of classifying thoughts or judgements—I return to this topic presently—is misguided, and analytic judgements, to speak in Kant's jargon, are not just *Erläuterungsurtheile*, sometimes they are *Erweiterungsurtheile* (Kant 1996, B11).[46] This is a philosophical criticism, not a scientific one.

I would like to add three short clarifying observations. The first is that Frege from the very beginning rejects the traditional way of classifying judgements:

> People distinguish *universal* and *particular* judgements: this is really not a distinction between judgements, but between contents. *They should say*, "*a judgement with a universal content*", "*a judgement with a particular content*". These properties belong to the content even when it is put forth, not as a judgement, but as an [unasserted] proposition [...]. The distinction of categorical, hypothetical, and disjunctive judgements appears to me to have only a grammatical significance (Frege 1879a/1972 §4; italics original).

Since the tradition did not separate sharply enough between content and judgement, they tended to talk about kinds of judgements, when they should have talked about content. The Begriffsschrift makes this evident. For as soon as the traditional subject-predicate logic is replaced by the argument-function logic, it is clear that the copula goes with the function and has no unique assertoric role (Frege 1906/1979, 185). The distinction between the categorical, hypothetical, and disjunctive, concerns the expression of content only, since, given a sign for negation, the hypothetical and the disjunctive forms are inter-definable (Frege 1879a/1972, § 7).

The second observation is that Frege would certainly object to Kant's apparent classification of both judgements and propositions [*Sätze*] as either analytic or synthetic (Kant 1996, B11–18). From Frege's perspective, the notions mark the kind of justification that could be given for a judgement. No wonder, then, that Frege states that, "I use the word 'thought' in roughly the same way as logicians use 'judgement'" (Frege 1906b/1979, 185).[47] The present understanding of the epistemological role of the Kantian notions in Frege's thinking is reflected by a remark from the introduction to *Grundgesetze*, where he formulates his project as that of determining once and for all the "epistemological nature" [*erkenntnistheoretischen Natur*] of arithmetic (Frege 1893/1967, vii.).

This takes us to the third observation. Until about 1892, Frege employed the words "analytic", "synthetic", "*a priori*" and "*a posteriori*", but thereafter classified judgements in terms of their source of knowledge [*Erkenntnisquelle*], as in "On Mr. Peano's Begriffsschrift and My Own" from 1897 (Frege 1897a/1984, 235).[48] There are three

---

[46] I use the German terms since Frege's play on words is lost in the translation.

[47] Cf. Frege (1918/1984, 354(n)). For informative historical-systematic accounts of the use of "proposition/thought" and "judgement" in logic, cf. Sundholm (2009) and Mahr (2010).

[48] On this very page, the plural "Erkenntnisquellen" is translated as "springs of knowledge", and the singular "Erkenntnisquelle" as "source of cognition". This slide between "knowledge" and

such sources, sense perception, the logical source of knowledge, and the geometrical and temporal sources of knowledge (Frege 1925/1979, 267). Towards the end of his life, Frege uses both the new terminology and the traditional Kantian one. I take this to be just a matter of terminology.[49] It might well be that the underlying reason for this change of expressions is that Frege realised that by using the traditional Kantian terminology, he encouraged a strongly Kantian reading of his philosophy, and that to express his views by the locution of sources of knowledge instead is a better way of capturing a basically epistemological project.[50] By considering this series of notions, repeatedly invoked by Frege: *Erkenntnistheorie*, *Erkenntnistheoretischen Natur*, *Erkenntniswert* and *Erkenntnisquelle*, the notational change strikes me as a wise move towards a uniform and telling terminology. Speculations aside, to say that a judgement is analytic, and to say that it is based on the logical source of knowledge, are, according to the present interpretation, nothing but alternative phrasings.

3.4 Judgements and Logic as Substantive Science

We have seen that logic is one of the sciences, and that science is characterised by judgements. Our final question is whether this feature of Frege's unquestioned starting point is in tension with his understanding of the axioms or primitive truths of the most general systematic science.

Let us look at Frege's phrasing of the judgement-stroke in *Grundgesetze*:

> We have already said that in a mere equation there is as yet no assertion: "2+3 = 5" only designates a truth-value, without its being said which of the two it is. Again, if I wrote
> "(2+3=5)=(2=2)"
> and presupposed that we knew 2=2 to be the True, I still should not have asserted thereby that the sum of 2 and 3 is 5, rather I should only have designated the truth-value of "*2+3=5"'s denoting the same as "2=2"*. We therefore require another special sign to be able to assert something to be true. For this purpose I let the sign " ⊢" precede the name of the truth-value, so that for example in
> " ⊢2²=4",
> it is asserted that the square of 2 is 4. I distinguish the *judgment* from the *thought* in this way: by a *judgment* I understand the acknowledgement of the truth of a *thought*. The presentation in Begriffsschrift of a judgement by the use of the sign " ⊢" I call a *proposition [Satz] of Begriffsschrift* or briefly a *proposition [Satz]*. (Frege 1893/1967, § 5; italics original)

In opposition to the passages looked at earlier, no "movement" or "advance" is mentioned, just a distinction between a thought and the acknowledgement of its

---

"cognition" is characteristic of the standard translations of Frege's writings.

[49] Cf. also Weiner (2004, 120 (n6))

[50] My reasoning about Frege's use of the Kantian notions in his classification of judgements should have made it clear that Carl's worry (in the cited passage from Carl (1994, 147)), that an alternative rendering of epistemic value would mean that Frege changed his view after *Grundlagen*, is unwarranted.

truth. Such a distinction is needed in order to carry out proofs in the Begriffsschrift, and is unproblematic (the sign then precedes the theorems, including the axioms, and, in Frege's phrasing, marks the propositions of Begriffsschrift). Now, our alleged problem is this: To grasp the content of a truth that needs no proof is sufficient to judge that it is true. This is not something particular to the science of logic, it holds for geometry as well. Furthermore, logic, as Frege conceived it, consists in drawing inferences from a given set of judgements. Such inferences, however, can be made completely absent-mindedly, without any reflections about content: instantiate the axioms and the derived theorems according to the recipe provided in *Begriffsschrift* or *Grundgesetze*, and apply modus ponens. As long as this is done correctly, one has derived theorems, that is, judgements. Frege himself uses the somewhat strange-sounding phrase "deriving judgements" [*Urtheile herzuleiten*] (Frege 1879a/1972, § 22). In such a case, the inwardly recognition of truth is absent. The fact that one can *derive* judgements mechanically appears to be a problem. To know that one has derived a truth is something different from making a judgement to the effect that a certain thought is true.

However, maybe this is not a problem after all? We have seen that Frege distinguishes sharply between setting up a system by way of stipulations and elucidations, on the one hand, and working inside the system, on the other. Talk about a judgement as a movement or an advance from a thought to a truth-value belongs to the description of scientific activity; it singles out the goal of science by underscoring that neither thought nor true-value is sufficient, the act of judging is needed as well. I am arguing that different aspects of the notion of a judgement are at play and must be kept apart: A judgement is a primitive act, and must be characterised by way of metaphors, such as a movement or an advance; while the judgement-stroke singles out the propositions in the axiomatic systems of *Begriffsschrift* or *Grundgesetze*.[51]

The notion of judgement as the aim of science, and the notion of judgement as giving "a proposition of Begriffsschrift" hang together, as only a tiny fraction of the infinitely many derivable judgements of the Begriffsschrift have epistemic value. The creative logician—a scientist—is characterised by his or her ability to form "scientifically fruitful definitions" and to derive significant propositions of the Begriffsschrift (Frege 1881/1979, 34, 1884/1968, § 70).[52] To conclude, even if there might be a problem about the metaphorical characterisation of a judgement as an advance *from* a thought *to* a truth-value, this problem belongs to the level of elucidations and carries no substantial or metaphysical commitment.

Concerning the issue raised at the very beginning, we could say that Frege's logicism is a programme for developing an axiomatic system, a collection of judgements that cover all of arithmetic, and although all of science, including logic, consists of judgements, the characteristic feature of these judgements is that they

---

[51] Cf. also the discussion about the judgement-stroke in Martin Gustafsson's article (Chap. 6) in this volume.

[52] As argued by Juliet Floyd, since definitions proper belong to the system, the Begriffsschrift has a potentiality for an ever-growing vocabulary (Floyd 1998).

are all analytic.[53] This insight about the analytic status of the judgements is not itself scientific, but philosophical or epistemological. We have, that is, a mathematical-philosophical foundation for arithmetic; or rather, would have had, were it not for the falsity of Axiom V.

Acknowledgments This article is a modified version of the talk "Logic as Substantial Science", held at the Conference *Frege zwischen Dichtung und Wissenschaft*, University of Bergen, December 5, 2014. I am grateful to the audience and the organisers of the conference for putting together a stimulating group of scholars and for ensuring an excellent social atmosphere. Although I am thankful to all participants for comments and reflections, I would like to mention three of them in particular. Juliet Floyd gave a most useful response to my talk, and Gisela Bengtsson and Joan Weiner, in addition to verbal responses, wrote comprehensive and quite instructive comments. This improved my line of thought and made possible the transformation of my talk into an article. The editors Gisela Bengtsson and Simo Säätelä suggested several useful improvements during the writing of the final versions.

References

Alnes, J. H. (1999). Sense and Basic Law V in Frege's logicism. *Nordic Journal of Philosophical Logic, 4*, 1–30.
Alnes, J. H. (2013). Trekk ved Freges logisisme: Begrep, logisk objekt og Aksiom V. *Norsk filosofisk tidsskrift, 48*, 243–264.
Bar-Elli, G. (2010). Analyticity and justification in Frege. *Erkenntnis, 73*, 165–184.
Benacerraf, P. (1981). Frege: The last logicist. *Midwest Studies in Philosophy, 6*, 17–36. https://doi.org/10.1111/j.1475-4975.1981.tb00426.x.
Carl, W. (1994). *Frege's theory of sense and reference*. New York: Cambridge University Press.
Costreie, S. (2013). Frege's puzzle and arithmetical formalism: Putting things in context. *History and Philosophy of Logic, 34*, 207–224.
Dummett, M. (1981). *The interpretation of Frege's philosophy*. Cambridge, MA: Harvard University Press.
Dummett, M. (1991a). *Frege: Philosophy of mathematics*. Cambridge, MA: Harvard University Press.
Dummett, M. (1991b). *The logical basis of metaphysics*. London: Duckworth.
Eder, G. (2015). Frege's 'On the Foundations of Geometry' and axiomatic metatheory. *Mind, 125*, 5–40. https://doi.org/10.1093/Mind/fzv101.
Floyd, J. (1998). Frege, semantics, and the double definition stroke. In A. Biletzki & A. Matar (Eds.), *The story of analytic philosophy* (pp. 141–167). London/New York: Routledge.
Frege, G. (1879a/1972). *Begriffsschrift/Conceptual Notation*. In G. Frege, *Conceptual notation and related articles* (pp. 101–204). T. W. Bynum (Ed. & Trans.). Oxford: Oxford University Press.
Frege, G. (1879b/1972). Applications of the "Conceptual Notation". In G. Frege, *Conceptual notation and related articles* (pp. 204–209). T. W. Bynum (Ed. & Trans.). Oxford: Oxford University Press.

---

[53] *Grundlagen*, Part V. Conclusion, begins thus: "I hope I may claim in the present work to have made it probable that the laws of arithmetic are analytic judgements and consequently *a priori*. Arithmetic does become simply a development of logic, and every proposition of arithmetic a law of logic, albeit a derivative one" (Frege 1884/1968, § 87). One reason why logicism is probable, but not certain, is that this guarantee can be obtained only by way of an axiomatic system.

Frege, G. (1880?/1979). Logic. In G. Frege, *Posthumous writings* (pp. 1–9). H. Hermes, F. Kambartel, & F. Kaulback (Eds.). Oxford: Basil Blackwell.
Frege, G. (1881/1979). Boole's logical calculus and the concept-script. In G. Frege, *Posthumous writings* (pp. 9–47). H. Hermes, F. Kambartel, & F. Kaulback (Eds.). Oxford: Basil Blackwell.
Frege, G. (1882a/1979). Boole's logical formula-language and my concept-script. In G. Frege, *Posthumous writings* (pp. 47–53). H. Hermes, F. Kambartel, & F. Kaulback (Eds.). Oxford: Basil Blackwell.
Frege, G. (1882b/1972). On the scientific justification of a conceptual notation. In G. Frege, *Conceptual notation and related articles* (pp. 83–90). T. W. Bynum (Ed. & Trans.). Oxford: Oxford University Press.
Frege, G. (1883/1972). On the aim of the "Conceptual Notation". In G. Frege, *Conceptual notation and related articles* (pp. 90–101). T. W. Bynum (Ed. & Trans.). Oxford: Oxford University Press.
Frege, G. (1884/1968). *Die Grundlagen der Arithmetik/The foundations of arithmetic*. J. L. Austin (Trans). Oxford: Basil Blackwell.
Frege, G. (1885/1984). On formal theories of arithmetic. In G. Frege, *Collected papers on mathematics, logic, and philosophy* (pp. 112–122). B. McGuiness (Ed.). Oxford: Basil Blackwell.
Frege, G. (1891/1984). Function and concept. In G. Frege, *Collected papers on mathematics, logic, and philosophy* (pp. 137–157). B. McGuinness (Ed.). Oxford: Basil Blackwell.
Frege, G. (1892a/1984). Über Sinn und Bedeutung/on sense and meaning. In G. Frege, *Collected papers on mathematics, logic, and philosophy* (pp. 157–178). B. McGuinness (Ed.). Oxford: Basil Blackwell.
Frege, G. (1892b/1984). On concept and object. In G. Frege, *Collected papers on mathematics, logic, and philosophy* (pp.182–195). B. McGuinness (Ed.). Oxford: Basil Blackwell.
Frege, G. (1893/1967). *Grundgesetze der Arithmetik I*. In G. Frege, *Grundgesetze der Arithmetik*. Hildesheim: Georg Olms Verlagsbuchhandlung. Partly translated in *The basic laws of arithmetic*. M. Furth (Ed. & Trans.). Berkeley: University of California Press.
Frege, G. (1897a/1984). On Mr. Peano's conceptual notation and my own. In G. Frege, *Collected papers on mathematics, logic, and philosophy* (pp. 234–249). B. McGuinness (Ed.). Oxford: Basil Blackwell.
Frege, G. (1897b/1979). Logic. In G. Frege, *Posthumous writings* (pp. 126–152). H. Hermes, F. Kambartel, & F. Kaulback (Eds.). Oxford: Basil Blackwell.
Frege, G. (1903a/1952). *Grundgesetze der Arithmetic II*. In G. Frege, *Grundgesetze der Arithmetik*. Hildesheim: Georg Olms Verlagsbuchhandlung. Partly translated in *Translations from the philosophical writings of Gottlob Frege* (3rd ed., pp.139–225). P. Geach, & M. Black (Eds.). Totowa, New Jersey: Barnes and Noble.
Frege, G. (1903b/1984). On the foundations of geometry: First series. In G. Frege, *Collected papers on mathematics, logic, and philosophy* (pp. 273–285). B. McGuinness (Ed.). Oxford: Basil Blackwell.
Frege, G. (1904/1980). Letter to Russell. In G. Frege, *Philosophical and mathematical correspondence* (pp.160–167). G. Gabriel, H. Hermes, F. Kambartel, C. Thiel, & A. Veraart (Eds.). Oxford: Basil Blackwell.
Frege, G (1906a/1984). On the foundations of geometry: Second series. In G. Frege, *Collected papers on mathematics, logic, and philosophy* (pp. 293–341). B. McGuinness (Ed.). Oxford: Basil Blackwell.
Frege, G. (1906b/1979). Introduction to logic. In G. Frege, *Posthumous writings* (pp. 185–203). H. Hermes, F. Kambartel, & F. Kaulback (Eds.). Oxford: Basil Blackwell.
Frege, G. (1906c/1979). A brief survey of my logical doctrines. In G. Frege, *Posthumous writings* (pp. 197–203). H. Hermes, F. Kambartel, & F. Kaulback (Eds.). Oxford: Basil Blackwell.
Frege, G. (1914/1979). Logic in mathematics. In G. Frege, *Posthumous writings* (pp. 203–251). H. Hermes, F. Kambartel, & F. Kaulback (Eds.). Oxford: Basil Blackwell.

Frege, G. (1918/1984). Logical investigations. Part I: Thoughts. In G. Frege, *Collected papers on mathematics, logic, and philosophy* (pp. 351–373). B. McGuinness (Ed.). Oxford: Basil Blackwell.
Frege, G. (1919/1979). Notes for Ludwig Darmstaedter. In G. Frege, *Posthumous writings* (pp. 253–257). H. Hermes, F. Kambartel, & F. Kaulback (Eds.). Oxford: Basil Blackwell.
Frege, G. (1925/1979). Sources of knowledge and the mathematical natural sciences. In G. Frege, *Posthumous writings* (pp. 267–275). H. Hermes, F. Kambartel, & F. Kaulback (Eds.). Oxford: Basil Blackwell.
Frege, G. (1926/1984). Logical investigations. Part III: Compound thoughts. In G. Frege, *Collected papers on mathematics, logic, and philosophy* (pp. 390–406). B. McGuiness (Ed.). Oxford: Basil Blackwell.
Heck, R. (2010). Frege and semantics. In M. Potter & T. Ricketts (Eds.), *The Cambridge companion to Frege* (pp. 342–379). New York: Cambridge University Press.
Jeshion, R. (2001). Frege's notions of self-evidence. *Mind, 110*, 937–976.
Jeshion, R. (2004). Frege's evidence for self-evidence. *Mind, 113*, 131–138.
Kant, I. (1996). *Critique of pure reason*. Indianapolis/Cambridge: Hackett Publishing Company.
Kitcher, P. (1979). Frege's epistemology. *Philosophical Review, 88*, 333–355.
Linnebo, Ø. (2003). Frege's conception of logic: From Kant to Grundgesetze. *Manuscrito, 26*, 235–252.
Linnebo, Ø. (2013). Freges oppfatning av logikk: Fra Kant til *Grundgesetze*. *Norsk filosofisk Tidsskrift, 48*, 219–230.
MacFarlane, J. (2002). Frege, Kant, and the logic in logicism. *The Philosophical Review, 111*, 25–65.
Mahr, B. (2010). On judgements and propositions. *Electronic Communications of the EASST, 26*, 1–20. http://journal.ub.tu-berlin.de/index.php/eceasst/article/viewFile/347/358. Accessed 23 Mar 2017.
Michaëlis, C. (1880/1972). Review of Frege's *Conceptual Notation*. In G. Frege, *Conceptual notation and related articles* (pp. 212–219). T. W. Bynum (Ed. & Trans.). Oxford: Oxford University Press.
Ricketts, T. (1985). Frege, the *Tractatus*, and the logocentric predicament. *Noûs, 9*, 3–15.
Ricketts, T. (1996). Logic and truths in Frege: I. *Proceedings of the Aristotelian Society, Supplementary Volume, 70*, 121–140.
Ricketts, T. (2010). Concepts, objects and the context principle. In M. Potter & T. Ricketts (Eds.), *The Cambridge companion to Frege* (pp. 149–220). New York: Cambridge University Press.
Salmon, N. (1986). *Frege's puzzle*. Cambridge, MA: The MIT Press.
Schröder, E. (1880/1972). Review of Frege's *Conceptual Notation*. In G. Frege, *Conceptual notation and related articles* (pp. 218–232). T. W. Bynum (Ed. & Trans.). Oxford: Oxford University Press.
Shapiro, S. (2009). We hold these truths to be self-evident: But what do we mean by that? *The Review of Symbolic Logic, 2*, 175–207.
Sundholm, G. (2009). A century of judgement and inference: 1837–1936. Some strands in the development of logic. In L. Haaparanta (Ed.), *The development of modern logic* (pp. 263–318). Oxford: Oxford University Press.
Weiner, J. (1990). *Frege in perspective*. Ithaca/London: Cornell University Press.
Weiner, J. (2004). What was Frege trying to prove? A Response to Jeshion. *Mind, 112*, 115–129.
Weiner, J. (2010). Understanding Frege's project. In M. Potter & T. Ricketts (Eds.), *The Cambridge companion to Frege* (pp. 32–63). New York: Cambridge University Press.

Chapter 4
Frege, the Normativity of Logic, and the Kantian Tradition

Anssi Korhonen

Abstract This paper considers the role of constitutivity and normativity in Frege's conception of logic. It outlines an historical interpretation with two goals. First, it traces these concepts back to their origins in Kant's philosophy. Second, it considers some of the different ways in which the issue of normativity and its proper grounding was addressed in the neo-Kantian tradition and in early analytic philosophy. Some neo-Kantians worked out an epistemic-normative conception of objective judgment, according to which the objectivity of cognition is constituted by distinctively logical norms. In Frege we find an original and sophisticated version of this line of thought. For Frege, the normative and constitutive roles of logic come to the fore in the articulation of scientific reason which follows the classical model of demonstrative science as *cognitio ex principiis* (cognition from principles). Wittgenstein's *Tractatus* then opens up a fresh Kantian perspective on the constitutivity of logic, one that grounds logic in structure rather than norms, and does so in conscious opposition to Frege and his normative science. Logic is transcendental, according to Wittgenstein, being the essence of the world and of all description. Hence, the normative function of logic becomes, in a way, superfluous.

Keywords Constitutivity • Logic • Normativity • Frege • Neo-Kantianism • Rationality • Transcendental deduction

4.1 Introduction

What is the relationship that logic bears to thought, or thinking, according to Frege?

Perusing the secondary literature, one can discern two lines of interpretation here.[1] On the first one, the relationship is seen as straightforwardly *normative*. This reading is apparently readily confirmed by Frege's own words. In the Introduction to

---

[1] Mezzadri (2015) offers a useful overview here.

A. Korhonen (✉)
University of Helsinki, Helsinki, Finland
e-mail: anssi.korhonen@helsinki.fi

Basic Laws, for instance, Frege argues that the laws of logic can be described as "laws of thought" in the sense that they state or prescribe (*festsetzen*) how one ought to think (Frege 1893/2013, xv).[2] Acknowledging that this is what he says, a careful interpreter adds that it would nevertheless be wrong to describe Frege as having endorsed a distinctively prescriptive conception of logic. Rather, that a law of logic has prescriptive force over our thinking is just a corollary of what the law is in itself; the laws of logic, Frege maintained, are the most general laws of being true, and it is for that reason that they yield norms that we should follow if we want to attain truth.[3]

According to the second interpretation, Frege regarded the relationship between logic and thought as *constitutive* rather than (straightforwardly) normative. On the constitutive reading, logic supplies rules without which there would be no thinking at all, somewhat in the way that the rules of chess are constitutive of the possibility of playing the game. To support the constitutive reading, the reference is made to passages where Frege argues, or seems to argue, that if one denies a law of logic, complete confusion ensues, so much so that thinking is no longer possible.[4]

Sometimes it is argued that constitutivity and normativity are exclusive of each other. After all, this seems to follow directly from what we mean by "normative rules" and "constitutive rules". Normativity presupposes the possibility of error, so that one's actions can accord with and fail to accord with a rule for that kind of activity. In the case of logic, this means that one has to be able make a logical mistake, as when one draws a conclusion that is not a logical consequence of the premises. On the normative conception, then, the laws of logic sort out episodes of thinking into those that are logically good and those that are logically bad; that is, there is such a thing as illogical thinking. If, on the other hand, the rules of logic are constitutive of thought, that supposedly means that there is no thinking in the absence of those rules, and hence there is no illogical thought. Apparently, the conflict between normativity and constitutivity is immediate.

Further inspection suggests, however, that the normative and constitutive conceptions of the laws of logic need not be in conflict. And some scholars have argued that Frege's understanding of the status of the laws of logic in fact combines normativity and constitutivity. For instance, MacFarlane (2002) argues that Frege's characterization of the laws of logic as the most general laws is primarily constitutive-normative, rather than descriptive: "Logic is general in the sense that it provides constitutive norms for thought as such, regardless of its subject matter" (MacFarlane 2002, 35). Similarly, Taschek (2008) explores the claim which Frege advances in the posthumous piece "My Basic Logical Insights" (Frege 1915) that the essence of logic is located in the assertoric force with which a sentence is uttered. This indicates, according to Taschek, that Frege took the laws of logic to be distinctively normative in that they "issue in constitutive norms of thinking as such" (2008, 384). Both MacFarlane and Taschek stress that constitutivity in no way rules out

---

[2] I follow here the translation of *Grundgesetze* by Ebert and Rossberg (Frege 1893/2013).
[3] Frege (1897/1983, 139, 1893/2013, xv).
[4] These passages include Frege (1884, §§ 14 and 26) and Frege's discussion of psychologism in the Introduction to *Grundgesetze*.

illogical thinking. As they understand it, the idea of constitutivity is merely that for a mental act to count as thinking, it has to be assessable as correct or incorrect in the light of the laws of logic; it need not actually conform to the laws of logic, and hence illogical thinking is possible (MacFarlane 2002, 37). Another formulation of this same idea is that to qualify as a thinker, it's enough if a person acknowledges, or is appropriately responsive to, the strictures of logic (Taschek 2008: 384). One can commit logical mistakes and still count as a thinker, as long as one regards oneself as a participant in the practice of reasoning and acts accordingly (whatever that might involve). Consider, then, a typical constitutivity-claim: Cerbone (2000, 298) argues that seeing logic constitutively, as "a background for our making judgments at all", excludes the normative conception, on which logic bears a prescriptive relation to judging. The appropriate response to Cerbone would seem to be simply this: How can logic form a background for our making judgments, if not through a recognition by us that it bears prescriptively on that activity?

It would seem then that normativity and constitutivity are relatively easily reconciled. And there is certainly textual evidence in Frege's writings for the reconciliatory interpretation, as well as against the constitutive interpretation.[5] Nevertheless, the interpretative issue of how Frege's conception of logic accommodates the normative and the constitutive strands deserves further discussion, if only to put it in a wider, historical context. In this paper, I try to supply at least a part of this context, by considering constitutivity and normativity from a perspective that is, perhaps, somewhat different from the standard one.

4.2 Regulative and Constitutive Rules in Kant's First *Critique*

The contrast between the normative and the constitutive is usually discussed with reference to Searle's distinction between regulative and constitutive rules (Searle 1969, Chap. 2.5). For the present purposes, however, it is more useful to consider the context in which it was originally introduced and which is to be found in Kant's philosophy.

Kant has more than one use for the distinction between constitutive and regulative rules (principles, judgments), but what is relevant here are the *synthetic principles of pure understanding* discussed in the Analytic of principles -section of Transcendental analytic. There are four such principles, namely Axioms of intuition, Anticipations of perception, Analogies of experience, and Postulates of empirical thinking. The first two are *mathematical* principles, and they are concerned with the construction of appearances and hence with their inner constitution. The latter two, on the other hand, are *dynamical* principles that merely "bring the existence of appearances under rules *a priori*" (1781/1787/1998, A179/B221).

---

[5] See, again, Mezzadri (2015) for details.

Walsh (1975, 99) explains the function of the mathematical principles: they "tell us how mere intuition or empirical intuition in general comes about". Hence, he continues, "they have to do with what might be called the internal structure of whatever falls within experience, the conditions which any experiential content must meet" (italics added). A further point may be added to this explanation: given that Kant's explanation of "content" goes hand in hand with that of "object", it follows that the rules underwritten by mathematical principles are rules for the generation of appearances of objects.

By contrast, the dynamical principles—for instance, the generic causal principle—do not concern the internal structure of appearances. Rather, dynamical principles are at play where the "occurrence of one item in experience is thought on the strength of these principles to demand the existence of some other item" (Walsh 1975, 99). As Kant himself points out elsewhere, the dynamical principles, too, are constitutive in that they partake in the constitution of *experience*. And in that capacity they stand on the same side of the fence as the mathematical principles, and opposed to what Kant calls the principles of pure reason, which can have no empirical use whatsoever (Kant 1781/1787/1998, A664/B692). Yet, the manner in which the dynamical principles are constitutive is quite different from the mathematical ones: the latter are constitutive "in regard to intuition", while the former principles are merely regulative principles of intuition. The dynamical principles, in other words, "make possible *a priori* the concepts without which there is no experience" (*ibid.*). This, indeed, is Kant's most succinct formulation of the distinction that we are concerned with.

Kant maintains, furthermore, that these two kinds of principles differ in their respective grounds of certainty: the mathematical principles possess "intuitive certainty", while the dynamical principles have only "discursive certainty". Yet, Kant also maintains that both are capable of possessing "complete certainty". Nevertheless, a crucial difference remains between these two kinds of principles. This shows up in Kant's treatment of the different a priori representations and their respective justifications.

The justification of the categories (the pure concepts of the understanding) is given by the A- and B-versions of the transcendental deduction in the Transcendental analytic, where Kant explains the justification for the use of such concepts as *cause*. The same question concerning justification applies to space and time as well. There is, however, a sharp contrast between the two cases: the pure concepts of the understanding and the pure intuitions; and this contrast parallels the one between the dynamical and mathematical principles. In the case of space and time, deduction is supposed to be unproblematic, because they can be shown to be forms of sensible intuition. In the case of the pure concepts, on the other hand, it is not clear just from the nature of what these representations are that they must be forms of what can be given to us in experience. Something can be an object for us (an appearance) only if it accords with the a priori determinations of space and time; hence, these determinations apply necessarily to all appearances. Categories, by contrast, "speak of objects through predicates not of intuition and sensibility but of pure *a priori* thought", and thus "they relate to objects universally, that is, apart from all conditions of sensibility" (Kant 1781/1787/1998, A88/B120). There thus arises a fresh problem for this variety of pure representations: how can the subjective conditions of cognition be objectively valid? We can easily imagine that there should be appearances without cause and effect, for example, but we cannot imagine objects that are not in space or time, according to Kant.

Kant himself believed to have proved that the pure concepts of the understanding are necessary for there to be objects of experience. The fact remains that their justification is quite different from that of space and time. We can articulate the difference by saying that the justification of categories is irreducibly *deontic* or *normative*: the justification lies, in the end, in our *right* to think of our experience as possessing a certain character, namely, as experience that is of objects and their interrelations which are independent of us. In the case of pure intuitions, by contrast, deduction yields a result that is *alethically necessary*: an appearance, whatever else it may be, is something that complies with the conditions of sensibility; these conditions, therefore, apply to appearances with absolute necessity and strict universality.[6]

The discussion in this section is summed up in the following table, inspired by David Hyder's account of Kant' two transcendental deductions (Table 4.1):

The terminology of "structural" and "normative deduction" is taken over from Hyder (2013). The a priori determinations of space and time are grounded in the structural features of the space of representation, whereas the categories are injected onto it, as it were; hence, also, the difference between the mathematical (constitutive) and dynamical (regulative) principles that are founded on these two kinds of synthetic principle.

In addition to pure intuitions and the categories, there is a third variety of pure representations. Kant calls them *logical functions of judgment*, and we may call them *pure forms of judgment*. They are importantly different from the other two kinds, when it comes to their justification. They are the subject-matter of the science that Kant calls *pure general logic*. Being general, this science "abstracts from all content of the cognition of the understanding and of the difference of its objects, and has to do with nothing but the mere form of thinking" (Kant 1781/1787/1998, A54/B78). Since general logic concerns in this way the sphere of the understanding alone, it knows nothing of sensibility. Accordingly, its subject-matter cannot be a concern for transcendental deduction, because that has to do, precisely, with the relationship between the two faculties. Considered in themselves, then, the forms of judgment have no *objectifying* or *object-giving function*, and the question of justification just

Table 4.1 Summary of Kant's transcendental deductions

Science	Kind of pure representation	Kind of transcendental deduction	Kind of synthetic principle of pure understanding
Transcendental aesthetic	Pure intuitions (space and time)	Structural deduction	Constitutive principles
Transcendental Logic	Pure concepts of the understanding (categories)	Normative deduction	Regulative principles

---

[6] We will meet this notion again in the final section of this paper.

does not arise for them. However, even these pure forms of judgment *can* be given a transcendental role.[7] All actions of the understanding can be traced back to judgment, and all concepts are predicates of possible judgments (Kant 1781/1787/1998, A69/B94); hence logical concepts (the pure forms of judgments) can be regarded as being justified, if not in themselves, then at least for us, by a kind of "transcendental analysis". This would be the reverse of the "clue" that Kant uses in the Analytic to discover the pure concepts of the understanding. Thus, the pure forms of judgments do not need a justification, when considered in themselves, but they can be transcendentally deduced by connecting them with the normative deduction of the categories.

4.3 Normativity and Constitutivity in Neo-Kantianism

Kant's successors introduced dramatic changes into the picture of human cognition that Kant had drawn up. But as we will also see, there was a continuing interest in the issues which had occupied Kant.

There was a wide-spread agreement among Kant's successors that a sharp distinction between sensibility and the understanding could not be upheld. This rejection took a particularly interesting form in the new logic that emerged in the nineteenth century. A typical advocate of the new logic maintained the following two theses. First, all knowledge in general and all "certain knowledge" in particular operates with logical forms and hence there are no irreducibly non-logical forms of thought.[8] Second, since it is clear that thought has objects, it is also clear that pure forms of thought cannot be completely separated from objects.

With the emergence of the new logic, the question of justification of logic becomes acute in a way it had not been for Kant. There were at least two sources for this new concern. First, since pure logical forms must be taken to possess an objectifying function, the logic of these pure forms can no longer be regarded as analytic in Kant's narrow sense; logic as the study of the pure forms of thought is not the "brief and dry" science that Kant took it to be (1781/1787/1998, A54/B78). In particular, the new logic cannot be taken to be epistemically unproblematic; on the contrary, it must be conceived not only as a priori but also as *synthetic*.[9] That, then, would be one reason to extend Kant's original critical question to the sphere of "pure logic".

There is another source for the critical concern, not wholly unconnected with the first one. It has to do with the view that logic, too, has its own characteristic concepts or entities. And the concern was strengthened by a special construal of what these logical objects are. A number of logicians and philosophers argued that in order to combat various psychologistic, sensualistic and inductivist trends, one has to acknowledge a

---

[7] Cf. here Haaparanta (1988, Sec. 4).

[8] Frege formulates the point for the special case of arithmetic, when he states that "there is no such thing as a peculiarly arithmetical mode of inference that cannot be reduced to the general inference-modes of logic" (1885/1984, 113).

[9] The young Russell was explicit on this point; see Russell (1903; § 434)

specific realm of logical entities—in the German-speaking philosophical worlds these thinkers were known as advocates of *Logizismus* (Ziehen 1920, § 45). These entities are not subjective like the ideas or *Vorstellungen* that inhabit our individual minds; nor are they objective in the sense of being actual or acting upon us causally; but they are nevertheless objective in the minimal, negative sense of not being dependent for their existence or properties on what we happen to think about them.[10]

There thus arise, in the context of the new logic, two possibly distinct questions that we readily recognize as variants as well as descendants of Kant's critical questions. First, what guarantee have we that we have correctly identified the entities inhabiting the "logical realm", together with principles concerning these entities? (Or, what guarantee have we that we have correctly identified at least *some* of these entities and their governing principles? There need be no presumption of completeness here). Secondly, what guarantee have we that there is an appropriate connection between the pure forms of thought, construed now as inhabitants of a logical realm, and the realm of ordinary objects, objects regarding which we make judgments using these logical forms?

Our interpretative question now is: Where did Frege stand with respect to these developments? One way to forge a connection between him and Kant is outlined in Hyder (2013). According to Hyder, Kant's use of the two deductive strategies—the normative and the structural—in the justification of foundational physical principles had important successors in some key developments in nineteenth century German philosophy of science. Hyder focuses on two developments: non-Euclidean geometries and their relationship to kinematics, as in Helmholtz and his neo-Kantian critics; and the a priori justification of force laws, as in Hertz's version of the picture theory, which was motivated by an elimination of "occult" distance forces in favor of concealed masses and concealed motions. Hyder argues, furthermore, that the contrast between norm and structure figures in early analytic philosophy as well, including the contrast between Wittgenstein's semantic account of logical necessity in the *Tractatus* and Frege's (and Russell's) axiomatic approach to logic. One of Wittgenstein's problems was to find the ground for the difference between what is genuinely logically necessary and what is merely universal. And he claimed to have found that ground in the degenerate relation that the so-called propositions of logic bear to the representational space of all possible atomic states of affairs. Frege, by contrast, conceived of logical propositions in "genuinely despotic terms" (Hyder 2013, 274), as flowing from the *laws of thought*. Here the necessity of logic is regarded as (genuinely) normative or deontic, and not as merely psychological, or anthropological; but unlike the *Tractatus*, Frege is incapable of providing a "justification of the binding character of these laws" (Hyder 2013, p. 275); the despotic-normative approach, that is, leaves open whether the laws it sanctions are actually true.

I find Hyder's general interpretation highly suggestive. In what follows, I will apply it somewhat differently from Hyder, however. I will consider Frege's concep-

---

[10] Frege is often included among "logicists" in Ziehen's sense, on the strength of his admonition that we must acknowledge a "third realm" (Frege 1893/2013, xviii–xix, 1918/1984). This issue will be examined in the later sections of the present paper.

tion of logic from the point of view of the concept of normative deduction, focusing on the broader issue of *judgment* and its *objective validity*. To bring out the contrast between norm and structure, I will then consider what I take to be the gist of Wittgenstein's neo-Kantian-of-sorts strategy of validating logic.

4.4 Kant and Frege on Objective Judgment

In the broad philosophical tradition of which Frege, too, was a member, the philosophical study of cognition and of knowledge is the study of judgment as something objectively valid: a judgment in the proper sense is *of objects* and possesses a *truth-value* in a manner which transcends mere subjective "taking-to-be-true". It is natural to associate these claims with Kant or Kantianism, broadly conceived. But one can be concerned with such issues as objectivity, genuine reference to objects, and possibility of truth and falsehood, and be in that broad sense "Kantian", without returning particularly Kantian answers to the problems raised by these issues.[11]

There is, on the face of it, a deep difference between Kant's and Frege's treatment of judgment, for Frege's account seems to be free of all *subjective* elements. As Tyler Burge observes, Frege thinks of judgments as "idealized abstractions, commitments of logic or other sciences, not the acts of individuals" (Burge 2000/2005, 12/357). On this view, "[i]ndividuals can instantiate these judgments through their acts of judgment, but the abstract judgments themselves seem to be independent of individual mental acts" (Burge 2000/2005, 12/357). Kant's account of judgment, on the other hand, is replete with reference to *representation, cognition, construction, synthesis,* and suchlike. Accordingly, in Kant's story about objective validity, the emphasis is apparently on the *mechanisms* that supposedly underlie objectively valid judgment, or indeed, objectively valid *judging*.

This gives Kant's notion of judgment its *subjectivist* character. And as historical evidence demonstrates, it makes Kant vulnerable to the charge of psychologism. The standard reply is to insist that the relevant mechanisms belong to *transcendental* and not to empirical psychology. This is fine as a reminder of what Kant saw himself as doing, but it is not as such very useful, as it does not explain what the connection is supposed to be between the mechanisms and their product.

The following perspective is more useful here. Consider one of Kant's more succinct expositions of objective validity, which is found in sections 18–20 of *Prolegomena*. There Kant explains that a *judgment of experience* arises when representations received through sensibility are brought together in an act of judging, and in such a way that they are connected "in consciousness generally", rather than "in a particular state of my consciousness" (Kant 1783/1902, § 20). The latter sort of connection yields no more than a *judgment of perception*, which is of subjective validity only and has no "reference to the object". This implies, among other things,

---

[11] F. H. Bradley's theory of judgment, as developed in Bradley (1883), offers a good illustration here.

that the traditional account of judgment as arising from *compositio et divisio*, or "putting together and taking apart", is mistaken, as it does not yield that "universality and necessity for which alone judgment can be objectively valid". A footnote appended to § 20 gives an illustration of Kant's meaning. Consider the judgment of perception, "When the sun shines on the stone, it grows warm". Here a number of perceptions are conjoined subjectively, and however often there occurs a perception of the sunshine warming a stone, the judgment does not contain necessity. It is only when the understanding's concept of *cause* is brought to bear on these perceptions, thereby connecting with the concept of sunshine that of heat as its necessary consequence, that there arises a judgment that is objectively valid. Such a judgment is one that possesses "necessary universality for everybody" (§ 19). The objective validity of a judgment involves the notion that "[w]hat experience teaches me under certain circumstances, it must always teach me and everybody" (§ 19). It is only in this way that a judgment "expresses not merely a reference of our perception to a subject, but a quality of the object"; for it is only if we assume that judgments refer to the unity of the object that we have a reason to think that other people's judgments are in necessary agreement with ours (§ 18).

Even this explanation of objective validity is not fully free of elements that someone might find subjective. Putting that aside, though, consider the core of Kant's explanation: judgments of experience arise when representations are brought together or connected *in consciousness generally*. This is to be distinguished sharply from the kind of connection that amounts to no more than a particular state of my consciousness. Furthermore, this consciousness generally is none other than what Kant elsewhere calls the *transcendental unity of apperception*, a notion that reintroduces the thinking subject. As Kant explains, "[t]he representation *I am*, which expresses the consciousness that can accompany all thinking, is that which immediately includes the existence of a subject in itself, but not yet any *cognition* of it, thus not empirical cognition i.e., experience [...]" (Kant 1781/1787/1998, B277). Thus it is also clear that the unity in question is of the most abstract kind: "The logical form of all judgments consists in the objective unity of apperception of the concepts contained therein"—which is the title of § 19 of the B-version of the transcendental deduction (Kant 1781/1787/1998, B140).

According to Kant, then, bringing representations together in consciousness generally is the same thing as judging by means of those representations. Furthermore, the nature of this activity is best brought out by means of a contrast. It involves putting forth something with a claim to being true or false in a way that is more than mere subjective conviction. Even this subjective conviction deserves to be called judging (at least according to the exposition given in *Prolegomena* and summarized above). Perhaps this is because it, too, involves a unity of sorts, albeit one that pertains only to a particular state of consciousness. A subjective unity in this sense is very close to what Kant elsewhere calls a "taking-to-be-true" (*Fürwahrhalten*); this "is an occurrence in our understanding that may rest on objective grounds, but [it] also requires subjective causes in the mind of him who judges" (1781/1787/1998, A820/B848). What emerges from Kant's discussion in sections 18–20 of *Prolegomena* is thus a distinction between judgments possessing a genuine claim to being true, on the one hand, and mere subjective conviction or takings-to-be-true, on the other.

When we interpret Frege from the perspective opened up by the concept of objective validity, we should align it precisely with sections 18–20 of *Prolegomena*. Such an interpretation sees him as operating with an *essentially logico-epistemic conception of objectivity*. An illuminating way of elaborating this reading is by comparing it with an *epistemic-normative* interpretation of Frege's conception of judgment. Its starting point is the generic claim that for Frege the notion of judgment is central or primary.[12] This claim is often made to emphasize the alleged distance between Frege and any approach (termed "Platonism" or "quasi-Platonism") that explains the objectivity of a priori cognition by postulating *objects* endowed with objectivity-conferring characteristics, such as independence of cognition, abstractness, immutability, and so on.[13] The *epistemic-normative* reading, though, is more specific regarding *which* feature of judgment it is that is supposed to constitute the ultimate source of objectivity for Frege; the guiding idea is, as Friedman (1992a, 536) puts it, that "the rules of logic make the notion of objective judgment or assertion possible in the first place". The essence of the epistemic-normative interpretation is thus that *logic has an objectifying function in its normative capacity*. Objectivity lodges in intersubjective agreement and disagreement. It is essential to my judgments and assertions that they can contradict and agree with yours and thus be subject to rational support and revision. These, in turn, are first made available by the general laws of logic, which define the relevant notions of agreement and disagreement and supply rules for rational argumentation. It is only in this way that judgment can be something more than a subjective conscious state, something that makes a genuine claim to truth and is therefore explicable in terms of reasons rather than causes.

An important source of inspiration for the epistemic-normative reading in contemporary scholarship on Frege is Thomas G. Ricketts' paper, "Objectivity and Objecthood: Frege's Metaphysics of Judgment". Ricketts starts from the uncontroversial point that Frege's notion of judgment centers on an absolutely sharp separation of an act of judging and a thought as its content. Where Ricketts differs from many traditional readings is in his claim that the separation is not meant to provide an *ontological* underpinning for objectivity. Rather, according to Ricketts, it involves a systematic redescription of "selected features of our linguistic practices", namely those that collectively "elucidate" the notion of objectivity (Ricketts 1986, 72). Once the focus is put on the notion of linguistic practice, objectivity becomes a matter of *intersubjectivity*, of genuine communication as opposed to an expression of subjective states or feelings ("ventings"). Crucially, furthermore, the redescription

---

[12] The generic observation is often backed up by citing such passages as Frege (1880–1/1983, 16–17, 1882/1983, 101) as well as Frege's letter to Anton Marty (Frege 1976, 163–165). We should note, however, that in all these passages Frege is in fact concerned with *concept-formation*. Frege's point is that traditional logic considers concepts as formed by abstraction, whereas he himself starts from judgments and their content. It is by no means straightforward to spell out the connection that this doctrine is supposed to bear on the issue of objectivity.

[13] Reck (2007) explores the contrast between the judgment-based approach and Platonism in detail and applies it to Frege.

is essentially *normative*. Thus, the act-content distinction is enforced by our appreciation of the relationship between yes-no questions and their answers. A yes-no question and an answer to it contain the same thought, but the accompanying acts are different: in the former case, it is a request for response; in the latter case, it is an assertion as a response to the request, namely, an acceptance or rejection of the thought (acknowledging it or its negation as true, as Frege would have it).

The opposition involved in assertion, Ricketts emphasizes, is to be sharply distinguished from the psychological impossibility of simultaneously accepting and rejecting a thought: the psychological impossibility and the logical opposition are kept separate from each other by locating the former in the empirical realm and the latter in *understanding*. A content, on this view, presents itself as something *judgeable*: it imposes a standard of correctness towards which one is logically obliged to take a stance, and that involves appreciating, among other things, the impermissibility of affirming downright contradictions (Ricketts 1986, 72–73). These standards, furthermore, are "inescapably applicable to any judgment; we cannot opt out of them and still take ourselves to be making judgments" (1986, 73). The relationship of logic to judgment, in other words, is at once *constitutive* and *normative*: the components that are to be distinguished in any judgment—a content, on the one hand, and its acceptance or rejection, on the other—are identifiable as such only through a series of appropriate "mays" and "musts".[14]

The suggested line of thought must now be elaborated.[15] Essentially, this will be a matter of relating Frege and his account of objective judgment not only to Kant but also, and crucially, to certain developments in neo-Kantianism.

4.5 Objective Judgment vs. Psychological Process

Kant is explicit that judgments, be they mere judgments of perception or judgments of experience, operate with *representations* (*Vorstellungen*). The difference between the two kinds of judgment consists in how their constituent representations are "brought together". Either they are brought together subjectively in a particular consciousness, a process which Kant sees as partly causal. Or else they are brought together in consciousness generally; that is, they are thought in accordance with the logical forms of judgment. Thus, judging always operates with representations. In that way even an objective judgment involves a psychological process.

---

[14] It goes without saying that the points made in the text are only illustrative and are not meant to exhaust the notion of "understanding a content".

[15] Ricketts' exposition does not in fact address the similarity between Frege and Kant at all, beyond pointing out that Frege's animus toward naturalism and empiricism and his "corresponding sympathy with Leibniz and Frege" makes understandable that he should have taken judgment as the starting point for his philosophy (1986, 66). A brief outline of the Kantian connection is given by Friedman (1992a), who is "very sympathetic" to Ricketts' reading of Frege (1992a, 535).

With Frege's conception of judgment, on the other hand, there occurs a radical break with the philosophical tradition that conceives thinking in terms of "ideas" or "representations" as something psychological. Frege makes the point clearly and forcefully both in the manuscript "Logik" (Frege 1897/1983) and the late essay on thoughts (Frege 1918/1984). On Frege's understanding of that term, "representations" are entities belonging to the inner world. Typical examples of representations are the sense impressions which we have and by having which we see and hear, etc. the things that are outside of us. Like impressions, then, representations in general not only need an owner, but are essentially private. It would be absurd, Frege maintains, to speak about you and me sharing what is literally one and the same representation, as when we have visual impressions when we watch one and the same green field (1918/1984, 67/361). Thinking and judging, on the other hand, involve *thoughts*. Thoughts are contents which are intersubjectively accessible, as when you take in with ease a complicated logical theorem that I manage to grasp only after a protracted effort. A thought, then, cannot belong to the content of a consciousness and nobody is its owner.

Intersubjectivity is not the only consideration that Frege uses to establish the point that thoughts are not representations. The same point emerges from various considerations relating to *truth*. Frege (1897/1983, 142) points out that a representation, being a picture, can be true only if it succeeds in fulfilling its purpose, as when it is asked whether this picture represents the Cologne Cathedral. If the purpose of representing something is lacking, there can be no talk of a picture being true. And this shows, Frege contends, that the predicate "true" does not in fact belong to a representation but to the *thought* that the representation depicts a certain object. In general, Frege holds that thoughts cannot be representations, since representations are neither intersubjectively accessible nor, properly speaking, such as can be characterized as true or false.

There is thus a difference between Kant and Frege: Kant did invoke *Vorstellungen* in his expositions of thinking and judging, while Frege did not. Beneath this difference, however, there is a deeper similarity between the two thinkers. That the constituents of Kant's judgments are "representations" does not reduce judgments of experience to purely psychological episodes or processes; this, after all, is one of the main lessons that we should derive from Kant's account of objective judgment in *Prolegomena*. Furthermore, Frege himself recognized and made this point very clearly in his earliest extended discussion of "subjective" and "objective". This discussion is found in sections 26 and 27 of *The Foundations of Arithmetic*. In a footnote to § 27, Frege draws a distinction between subjective and objective representations. The former are governed by the psychological laws of association and are "of sensible, pictorial character", whereas the latter are non-sensible, belong to logic and can be divided into objects and concepts. He then adds the interesting remark that Kant linked both meanings to his use of *Vorstellung*, thereby "giving his doctrine a very subjective and idealist coloring", and rendering his true meaning difficult to discover.

In § 26, Frege has just distinguished between what is subjective and what is objective. The subjective means dependence of sensing, intuiting, representing and constructing mental images; the objective involves reason (*Vernunft*) and, therefore, judgment. Frege now aligns Kant's use of "representation", or what he regards as Kant's true understanding of the concept, with the latter; so, Frege saw himself and Kant standing on the same side with respect to the present issue. True enough, Frege's mature doctrine of judgment, with its notion of thought, is not yet there in *The Foundations*. But the separation of the objective from the subjective is effected in essentially the same way in 1884 as it would later be effected; for Frege, it always involved psychological processes on the subjective side and judgment and reason on the objective side. The essential point here is that Frege attributes this very distinction to Kant as well—although he also argues that Kant was not as clear about the matter as he should have been.

In Frege's mature theory of judgment, the contrast between what is objective and what is subjective goes hand in hand with an absolutely sharp distinction between thoughts and representations "in the psychological sense of the word" (1897/1983, 146). This does not yet imply that think*ing* and judg*ing* could not be psychological processes. And Frege appears to think that this is what they are. For he states that "[w]hen we *inwardly* acknowledge something as true, we make a judgment" (1897/1983, 150; emphasis added), a characterization which goes together with that of assertion as the "manifestation (*Kundgebung*) of [a] judgment" (1918/1984, 62/356). Furthermore, his well-known dictum says that when we think, we *grasp* thoughts, and he is explicit in calling this grasping a "mental process" (*seelischer Vorgang*) (1897/1983, 157).

Two caveats should nevertheless be added here. Firstly, Frege argues that grasping is something which already lies on the border of the mental and which, for that reason, cannot be fully understood from a psychological point of view: "[…] in the process, something comes into view that is no longer mental in the proper sense: the thought; and this process is perhaps the most mysterious of all" (1897/1983, 157). Secondly, he puts aside the whole issue of grasping, maintaining that in logic we need not trouble ourselves about such mysterious processes. There it is enough that we do, in fact, grasp thoughts and acknowledge them as true.

In the present discussion, however, the notion of grasping cannot be put aside. The epistemic-normative interpretation argues in general that objectivity is tied to objective judgment, and the objectivity of judgment lodges in the normative force of the underlying logical rules. Given this, the question, "What is involved in this 'grasping' and, more generally, in the understanding of content?" becomes inextricably intertwined with another: "What is involved in our appreciation of the normative force that logic has on judgment?" If, as was suggested in Sect. 4.1, logic is constitutive for judgment through a recognition that it bears prescriptively on judging, then the question arises: What form does this recognition take? And behind this, there is another question: What is the *ground* of the normative force of the underlying logical rules?

4.6 Norms and Their Ground

Hyder (2013) made a twofold interpretative claim: first, that Frege's justificational strategy for his basic laws builds on the distinction between genuinely normative laws and laws that are merely empirical or psychological; and second, that this strategy is to be aligned with Kant's normative deduction. Hyder claims, furthermore, that Frege's strategy was an outgrowth of the *value-theoretic form of neo-Kantianism*, as exemplified by Hermann Lotze and Wilhelm Windelband (*ibid*. 275).[16]

This reference to neo-Kantianism adds an important new dimension to the notion of normative deduction. In Kant's case, the contrast between normative and structural deduction suggests an association of the former with *free-standing normativity*, as we might call it: in the case of the structural deduction, as we have seen, the validity of a concept is grounded in structural features of the space of representation; the validity of a normative rule, on the other hand, derives ultimately from *spontaneity*, from the *self-activity* of a faculty (Kant 1786/2002, 68 [Ak 4:452]).

Spontaneity admits of degrees. Famously, Kant uses spontaneity vs. receptivity to distinguish the understanding from sensibility (Kant 1781/1787/1998, A50–1/B74–5). In this case, however, spontaneity does involve a restriction. The function of spontaneity is to introduce order, by means of concepts that are "produced from its activity", into the manifold of sensibility, which is pre-given in some sense. When sensibility is put aside, on the other hand, what remains is *reason* describable as *pure self-activity*, "which, under the name of the ideas, shows such a pure spontaneity that it thereby goes far beyond everything that sensibility can provide it" (Kant 1786/2002, 68). Thus, the understanding, unlike reason, *is* constrained from without. But the constraint is not such as to enable a *derivation*, in any sense, of normative necessity from something more fundamental. On the contrary, the rules of the understanding are distinguished from sensibility by such free-standing normativity.[17]

In neo-Kantianism, the issues surrounding normativity become quite complicated. On the one hand, some neo-Kantians argued that in order to steer clear of

---

[16] Hyder does not elaborate on the second part, but refers to Gottfried Gabriel's work on Frege: see Gabriel (2005, 2013).

[17] On the other hand, when the relationship to the sensible manifold is abstracted away, what remains are the "rules of the understanding in general". These, too, are characterized by a kind of pure self-activity, but here this self-activity or spontaneity is not immediately related to normativity. Rather, such rules *define* or, better, *constitute* what the understanding is in itself: they are the "absolutely necessary rules of thought without which there can be no employment whatsoever of the understanding" (Kant 1781/1787/1998, A52/B76). The formal counterpart of spontaneity in this constitutive sense is *self-consistency*. As Kant explains in *Logik Jäsche*, the rules of general logic are supposed teach us the "*correct* use of the understanding, i.e., *that in which it agrees with itself*" (1800/1992, 14/529; emphasis added). Here talk of "correct use" does not, I think, refer to normativity (correct vs. incorrect use of something). Instead, it indicates that these rules are rules of *self-consistency*, or rules for the understanding's *self-agreement*. Normativity only arises when such rules are applied to actual cognition. The possibility of a logical error thus arises from the "unnoticed influence of sensibility" upon judgment; it arises when we confuse merely subjective grounds of judging for such as a genuinely objective (Kant 1800/1992, 53–54/560–561).

psychologism, a philosophical account of cognition has to draw on normativity, properly so called; this is also the guiding idea behind the epistemic-normative reading of Frege. On the other hand, other neo-Kantians, while thinking through what is involved in normativity, argued that a "derivation" or grounding of the normative was needed. And the complication here arises from the fact that the derivation was used to secure one or the other of two distinct goals: *either* the proper *normativity* of what is so derived *or* its *autonomy*.

Indeed, one could bring normativity and autonomy together.[18] According to Hermann Lotze's influential account, logical laws are characterized first and foremost by their *validity*, which exhibits just this duality.[19] Validity is a primitive kind of *reality*. As such, it is irreducible to anything that does not already involve validity. In particular, the laws of logic are independent of and autonomous from all psychological and other such contingent characteristics of judging subjects. At the same time, calling these laws *valid* is meant to emphasize their irreducibly *normative* status as well.

Philosophers influenced by Lotze separated these two lines of thought, thereby formulating two distinct argumentative strategies against psychologism.[20] Wilhelm Windelband, for instance, came to endorse the normative element in Lotze's notion of validity. Windelband argued, in his late *Die Prinzipien der Logik*, that we must draw a clear distinction between *validity in itself* (*Geltung an sich*) that belongs to a law of logic and its *validity for us* (*Geltung für uns*). The latter has its ground in the former: seen from our point of view, the logical (*das Logische*) is an "ought"; but this "ought" must have a ground in something that possesses validity in itself (*dessen Geltung an sich besteht*), and becoming an "ought" only through a relation to a consciousness that is capable of erring (Windelband 1912, 18).

Even this emphasis of normativity, however, seems to involve an element of independence, or autonomy. For what characterizes norms is their "immediate evidence for our consciousness" (Anderson 2005, 304, quoting from Windelband's *Präludien* of 1884). At least on Anderson's reading of him, this leads Windelband to accord the logical norms with a quasi-platonic status (Anderson 2005, 305).[21]

On a view like this, when one gives an account of objectivity, one is not forced to choose between normativity and "Platonism" (with or without a "quasi-"). The conceptual situation thus becomes quite complex. Returning to Frege, we can see that

---

[18] This third option seems to apply to Kant's notion of pure general logic, which does involve both autonomy and normativity. The rules of general logic, which are norms for actual thinking, are *autonomous* because they are the defining rules of the faculty of the understanding. And they are genuinely *normative because of this autonomy*.

[19] Here I follow the exposition of Lotze that is given in Anderson (2005, 294–6). See also Ziehen (1920, §§ 45–6), which distinguishes between two kinds of "logicism": one that emphasizes the independent "existence of specific logical entities" (this was mentioned above in Sect. 4.3); the "value-theoretical logicism", which decrees that all logical reflection is founded on the fundamental fact that we draw a value distinction between true and false representations.

[20] Again, I am here following Anderson (2005).

[21] Gottfried Gabriel, too, speaks of "transcendental Platonism" in this connection (Gabriel 2013, 287).

in interpreting him, it is not enough to point out a general similarity between his views on logic and those endorsed by neo-Kantians, or even by a particular strand of neo-Kantianism.

Such similarities can, indeed, be found. Gabriel (2013, 287–8) mentions two important points of contact between Frege and value-theoretic neo-Kantianism. First, there is an emphasis on the duality in the notion of a logical law. Frege (1893/2013, xv) points out that the word "law" has a "double meaning": in one sense a law asserts what is, whereas in another sense it legislates what ought to be. Similarly, Windelband (1912, 18) argues that the laws of logic "always have a two-fold nature": the laws of logic are rules for our thinking, or judging, but these laws have their nature quite independently of actual processes of representation and of our actual judgments. Second, the link between laws in the second sense and norms is established through a "will to truth": in Windelband's words, the laws of logic are "rules to which all thinking that is directed towards truth must conform" (1912, 18); in Frege's words, these laws "can be regarded as rules for judging with which it must comply, if it is not to miss the truth" (Frege 1897/1983, 157).

According to Gabriel (2013, 287), there can be no hermeneutical doubt that Frege belongs to the neo-Kantian tradition and, in particular, to its value-theoretical version. This interpretation, I will now argue, must at least be complemented by a consideration of what is in fact most distinctive in Frege's contribution to the explication of the triangle of objectivity, reason, and normativity. I would like to put the point as follows:

> For Frege, reason and, with it, objective judgment and normativity are not best captured by free-standing norms for judgment, as in Kant[22]; or even by a "derivation" of normativity from validity, as in some neo-Kantians; but by an articulation of what is involved in *scientific reason*.

On this understanding, reason is aligned, not just generally with judging as the acknowledgement of truth but more specifically with the *acquisition of knowledge*. This perspective surfaces in Frege's late essay *Die Verneinung* (Frege 1919/1984), where judgment and assertion are conceptualized as the end point of a strongly idealized process of inquiry, constituting *justification*. In one of Frege's last essays, we find the following formulation:

> A unit of knowledge comes into being when a thought is recognized as true. For that purpose, the thought must first be grasped. Yet I do not count the grasping of the thought as knowledge, but only the recognition of truth, judgment proper.[23]

The idea that logic is *constitutive* of judgment becomes quite intuitive when it is embedded in the larger context provided by the notion of justification; when logic is seen through the role that it has in a process of inquiry, it is natural to conceive of its rules as constitutive of the process. And it becomes even more natural when—as in

---

[22] Evidently, "reason" is here used in a sense that is broader than Kant's; it is the sense that Frege uses in his discussion of objectivity in sections 26 and 27 of *Grundlagen*.

[23] Frege (1924–25, 286; my translation).

Frege—the focus is on a priori knowledge and on proof as the proper form of justification for this type of knowledge.

For Frege, the notion of judgment as knowledge goes hand in hand with the *classical model of demonstrative science*. This is itself a model of scientific rationality,[24] and in Frege's hands it has two important consequences. First, it offers him a general framework for his thinking about logic and logical notions.[25] Second, it leads him to formulate logic itself as a theory the design of which follows, by and large, the classical model.[26] This is crucial for our discussion, as it opens up two perspectives on what is involved in the articulation of the scientific reason. On the one hand, logic comprehends what may be variously described as the universal norms for thought, the principles of correct reasoning, or the "set of universally constraining facts" (Kemp 1998, 222); it is these norms, principles or facts the recognition of which is seen as normatively constitutive for judging, that is, for the establishment of knowledge. On the other hand, there are also *particular formulations* of such principles, like Frege's own *Begriffsschrift*. Moreover, the second perspective should not be written off as possessing only marginal importance. On the contrary, insofar as one is concerned specifically with the articulation of scientific rationality, the second perspective is indispensable. This applies to Frege and his attempt to *prove* the logicist thesis, which was a thesis about the nature of arithmetical knowledge.

4.7 Frege and the Articulation of Reason

Kant argued that the forms of thought are essentially empty without the forms of sensibility. His point was not only about what we can and cannot experience, and hence not only about what forms of thought exhibit "real possibilities". It was primarily a *representation-theoretic* point, namely that the representational resources of traditional formal logic—what Kant called general logic—are insufficient for the representation of any genuine piece of knowledge. Not even judgments of pure mathematics, in spite of being a priori, can be as much *formulated* using logical

---

[24] Cf. de Jong and Betti (2010).

[25] This leads Frege to a broadly epistemic conception of logic, on which logic constitutes the method of justification. On such a view, the epistemic function is not somehow external to logic, a field for "applied logic". Rather, it is built into what logic is. This shows up in Frege's description of the very core of logic: "Logic is concerned only with such grounds of judgment as are truths. Judging by being aware of other truths as grounds of justification is known as *inferring*. There are laws for this kind of justification, and the goal of logic is to establish these laws of correct inference" (Frege 1879–1991/1983, 3; my translation, with emphasis in the original). On this view, the subject-matter of logic is not mere logical consequence but "correct" inference, or inference that is valid, the establishment of truths on the basis of other truths.

[26] "By and large", because Frege does not simply adopt the classical model of scientific knowledge as *cognitio ex principiis* (knowledge from principles; cf. de Jong and Betti (2010: 186) , but modifies it certain subtle ways; for a detailed discussion of this issue, see Burge (1998/2005).

forms alone.[27] What must be added to pure logical forms are the forms of intuition, and only in this way can there be mathematical judgments conveying genuine mathematical content.[28] Frege's discovery that the realm of pure logical forms far outreaches the scope that logical tradition and Kant had acknowledged put Frege in a position to dispense with intuition in this representation-theoretic sense.[29] In particular, he argued that judgments and proofs in mathematics do not borrow anything from intuition. If it looks otherwise, that is only because we have not taken the trouble to formulate the relevant stretch of mathematical knowledge in a logically perspicuous way (Frege 1884, § 90).

Now, one may argue that this difference between Kant and Frege has far-reaching consequences for their respective articulations of reason; the following summarizes Friedman (1992a). The way Kant draws the distinction between understanding and sensibility suggests that the former has a *wider scope* and that, therefore, the forms that mathematical cognition operates with—space and time—are necessarily less general than the categories. For when sensibility is abstracted away, what is left is the form of thought (categories) and, with it, the concept of an object in general. In a sense, then, categories "extend further" than intuition, because categories "think objects in general without seeing to the particular manner (sensibility) in which they might be given" (Kant 1781/1787/1998, A253–4/B309).

For Kant, the wider reach of the understanding secures a perspective from which we can meaningfully think without intuition. In particular, there are not only the understanding and sensibility but also the faculty of *reason*, construed now as a *meta-faculty* and the place where transcendental philosophy itself unfolds. Philosophy, in its Kantian-cum-critical function, is a meaningful and rational enterprise. And the project of transcendental deduction, i.e., the proof that there is a necessary harmony between sensibility and the understanding, can itself be given a theoretical justification within this meta-discipline of reason.

Frege's position is in this respect decisively different, according to Friedman. For Frege, thinking unfolds everywhere according to logical forms, and there can be no meta-faculty in which one could meaningfully and theoretically discuss those forms themselves or philosophize about them. What there is, is the logical decomposition of rational discourse, as effected in Frege's *Begriffsschrift*, which "at once embodies all the conditions of rationality and contains everything that can be thought—for every thought, regardless of its subject-matter, must have a determinate logical relation to every other thought" (Friedman 1992a, 538). It is this *universality* of logic—

---

[27] See Kant (1781/1787/1998, A162–163/B203; A234/B287; B154–155).

[28] The representation-theoretic reading of Kant is best known from Michael Friedman's work: see Friedman (1992b, Chapters 1 and 2).

[29] Frege did accept, though, that there is *a* link between geometrical knowledge and *Anschauung*. But this link is not needed to give geometrical thoughts mathematical *content* but to render them *true*, when they are true. This view show up in Frege's understanding of non-Euclidean geometries. As Burge (2005, 60) observes, Frege acknowledged the existence of non-Euclidean geometries, and held them to be (i) mathematical curiosities, (ii) consistent but (iii) false, because ruled out by *Ansschauung*. Therefore, we may add, non-Euclidean geometries are not knowledge in the proper sense delineated by the classical model of demonstrative science.

its omnipresence and unrestricted bearing on every discourse—that precludes Frege from as much as formulating the analogue of the problem of the transcendental deduction for his *Begriffsschrift*. In that sense, Friedman concludes, Frege's approach constitutes the very antithesis of genuine transcendental philosophy.

What emerges from Friedman's reading of Kant vs. Frege on the articulation of reason is a familiar picture of Frege as someone who was committed to a conception of logic and rationality whose implications become fully explicit only in the *Tractatus*. In that work, we find two key ideas: first, that the "conditions of possibility of all rational thought" (of representation, of all meaningful use of language) are to be found in Frege's new logic—or the sound core that can be gleaned from Frege's new logic; and second, that these conditions nevertheless cannot be "stated, formulated, argued for, or justified" (Friedman 1992a, 538). The result is a non-doctrinal conception of philosophy that Friedman detects in Frege, too. On this view, Frege was not concerned with developing a system of transcendental philosophy, which would comprise the conditions of rationality and objectivity; he was concerned with the rigorous formulation of logic with a view to using it for a specific scientific purpose (*ibid.* 539).

Friedman's reading of the Frege-Wittgenstein connection faces two major difficulties. First, it overlooks the difference between the two Kantian deductions. Second, it fails to differentiate sufficiently between universal logical rules or norms, on the one hand, and their explicit formulation in a theory (or calculus or system, etc.) of logic, on the other; or more generally, Friedman's reading does not observe the distinction between *logic in its basic function*, whatever that may be, and *logic as a theory* or *logic as a doctrine*. The distinction can be used for different interpretative purposes, but here I will use it to outline two points: (i) that what is distinctive in Frege's conception of logic is that he sees logic in its basic function as normatively constitutive for judging *qua* knowledge; (ii) that Wittgenstein, rather than following out Frege's conception of logic, in fact replaces it by one that differs radically from Frege in what it implies for the normativity of logic, among other things. One way to conceive this matter is to see how Frege does, while Wittgenstein does not, allow for a meta-perspective. The following table summarizes their main differences in this respect: The Table 4.2. "Main differences between Frege's and Wittgentein's conceptions of logic",

(Here "logic$_F$" refers to logic in the fundamental sense, the function of which is articulated in the second cell; "logic$_T$" refers to logic as theory or doctrine.) According to Frege, it falls to logic to discover, or discern, *the laws of truth*, or *the most general laws of being true* (1897/1983, 139). He likens the laws of logic to the

Table 4.2 Main differences between Frege's and Wittgenstein's conceptions of logic

Frege	Wittgenstein
There is not only logic$_F$ but also logic$_T$: meta-perspective is therefore possible	There is only logic$_F$: Meta-perspective is not possible
Logic$_F$ is normatively constitutive for judging *qua* knowledge	Logic$_F$ is the essence of all description/of the world: Logic$_F$ is *structural-transcendental*
Logic$_T$ is a *doctrine*: a theory of correct justification	Logic$_T$ does not exist: there is only *Begriffsschrift qua* perspicuous notation

laws of nature (Frege 1918/1984, 58/350). As such, they are what they are independently of what we take them to be. They come to possess a normative and constitutive role, and this happens for the reason indicated above: insofar as our concern is with knowledge and therefore with truth, we must observe the bearing of the laws of logic—that is, the laws of truth—on us, for otherwise there can be no question of justification and establishment of knowledge, properly so called.

Frege holds that the normative reach of logic extends to our actual or ordinary judgments. He expects everyone to agree that logic supplies "guiding principles for thought in the attainment of truth" (1893/2013, xv); any pursuit of truth is governed by the laws of logic, which prescribe "what ought to be judged, wherever, whenever, and by whomsoever the judgment may be made (*ibid*. xvii).[30] It is nevertheless clear, I think, that, for Frege, logic is normative and constitutive, first and foremost, in the function that it performs in the articulation of scientific reason. And this in fact means two things. First, $logic_F$ is involved in the scientific justification of any piece of knowledge (Frege 1884, § 3). Second, this presupposes that $logic_F$ itself be given an explicit formulation, what Frege needs is thus a "calculus" or "model" or "theory" of $logic_F$.[31]

Friedman' reading of Frege assigns two properties to Frege's *Begriffsschrift*: first, it "at once embodies all the conditions of rationality"; second, it "contains everything that can be thought" (Friedman 1992a, 538). For Frege, Friedman concludes, "there is no point of view from which we can meaningfully think independently of the *Begriffsschrift*", and hence that "there is no meta-perspective from which we can philosophize about the *Begriffsschrift*" (*ibid.*) Such conclusions are called into question, however, once we observe the distinction between $logic_F$ and $logic_T$. This gives us a good reason to maintain the following claims. First, there is in Frege's *Begriffsschrift* no presumption of "completeness" in Friedman's sense; if all conditions of rationality belong to logic, then $logic_F$ comprehends all such conditions; but *Begriffsschrift* is just $logic_T$. Second, saying that the *Begriffsschrift* qua $logic_T$ "embodies everything that can be thought" can only mean that every thought that is formulated in the *Begriffsschrift*, simply in virtue of being a thought, bears a determinate logical relation to every other thought. This is a consequence of the universality of logic that Frege accepted (or so I shall assume). So, the claim is about $logic_T$ only in the following sense: if logic supplies norms for thinking, they will be binding for all thinking; if so, they apply to judgments that one makes when one reasons about a $logic_T$, too.

---

[30] Frege emphasizes that this does not apply to the laws of logic only; for any law of nature stating what is the case is a law of truth and can, accordingly, be regarded as supplying a norm for judgment (1893/2013, xv).

[31] At least, this holds for the purposes that Frege has in mind: see Frege (1884, §§ 3–4, 1893/2013, vii). He is also sharply critical of colleagues like Dedekind, whom he sees as pursuing a line of investigation similar to his own but who does not make explicit the logic—i.e., $logic_T$—that he uses. Frege observes that Dedekind has managed, in his *Was Sind und Was Sollen die Zahlen*, to push the foundations of arithmetic much further, and in much less space, than Frege. But that is only because in Dedekind's book, "much is not in fact proven at all" (Frege 1893/2013, viii).

To be sure, given the conception of logic (and thinking) that we are examining here, it follows that there is no thought independently of logic$_F$. What follows? The consequence that Friedman has in mind clearly has to do with *justification* and its alleged non-availability to Frege. Friedman's point is that while Kant could provide, within the sphere of reason, a "deduction" for the application of categories to appearances, Frege was not in a position to offer an analogous justification for what he recognized as conditions of rationality. And if by this is meant that Frege did not think that he could give *inferential*, *non-circular*, and *persuasive* justifications for what he took to be principles of logic$_F$, that is, no doubt, completely correct. This conclusion presumably follows on anyone's conception of logic; and if one is a universalist about logic, it follows immediately.

But this does not render "justification" an otiose notion, when applied to logic. On the contrary, Frege had excellent general reasons, stemming from how he conceived the role of logic in the articulation of scientific reason, for being interested in the justification of his own logic$_T$: for it is only if the logic that he actually uses satisfies a number of criteria that it can be put to a good use in delivering scientific justification for arithmetical knowledge and more generally. To take an obvious example, the rules that one actually uses to derive conclusions from one's premises had better be valid; if not, the conclusions that one draws using the rules do not really follow, and the derivations fail to serve their epistemic function. Thus, when Frege argues in § 14 of *Grundgesetze* that his "first method of inference", which is a version of *modus ponens*, is truth-preserving, he is not thereby presenting a *validation* for a rule that belongs to logic$_F$. His point is to show that his particular version of that rule in fact has a certain desirable property, namely, that its use in the *Begriffsschrift* only ever yields true sentences as conclusions if one's premises are true sentences.

Judgment in Frege's sense involves justification and knowledge. Understood in this way, judgment would not exist without acknowledgement of the role of logic$_F$. Logic$_F$ is thus constitutively normative for judging as knowledge. Crucially, and unlike on the Friedman-style interpretation, Frege does not think that the normative-constitutive role of logic$_F$ has to remain unarticulated; on the contrary, he thinks that this role is to be spelled out by formulating an explicit theory of logic, and applying that theory in the articulation of actual pieces of scientific knowledge. Given Frege's perspective, furthermore, there is no reason to think that the acceptance of logic should have to be blind. Logic itself—that is, logic$_T$—is a doctrine, the formulation of which is a rational process, although the relationship between logic$_F$ and logic$_T$ gives that process a special shape and imposes limits on what it can achieve.

4.8 Wittgenstein and the Structural a Priori

In Wittgenstein's *Tractatus logico-philosophicus* (Wittgenstein 1921/1922/1961; I will refer to it as *TLP*), we encounter a conception of logic that completely rejects the idea that logic is a doctrine and the accompanying notion that logic is normative.

Points which are meant to undermine components of logic-as-doctrine are found throughout the book. For example, *TLP* 5.13 and comments thereon explain the nature of consequence and inference, dismissing as senseless and superfluous 'laws of inference' (*Schlussgesetze*), which supposedly justify inference. Again, logically primitive signs and their introduction are discussed in *TLP* 5.45 *et passim*, where Wittgenstein asserts that there are no privileged numbers in logic, no classification and no degrees of generality. The gist of the matter, however, comes in the 6.12's and 6.13. They explain that since the propositions of logic are tautologies and say nothing, they are all of equal rank; thus there is in logic no division of propositions into those that are essentially primitive and others that are derived from them.[32] Logic, therefore, is not a theory, a body of doctrine (*Lehre*); logic is transcendental, a mirror-image or a reflection (*Spiegelbild*) of the world (*TLP* 6.13).

Wittgenstein's alternative understanding of logic involves more than just an analogy of Kant's structural deduction. First, though, we need a reminder of what Kant actually did. His claim was that we can think of geometrical objects only by constructing their concepts and that this presupposes space and time as forms of sensibility.[33] Consider, for instance, the following proposition:

(AS) Necessarily, every triangle has the angle sum property.

The only way we can *think* of triangles and the angle sum property is by constructing that property—a construction whose mathematical presentation is Proposition I.32 of Euclid's *Elements*.

It follows that any object to which the concept *triangle* applies has the angle sum property. Accordingly, the concept *triangle whose interior angles do not add up to two right angles*, since this cannot be constructed, does not represent any genuine possibility, although it cannot be faulted by means of logic or conceptual analysis alone. Understood in this way, the modality in "cannot be constructed" becomes genuinely *alethic*, and not normative, let alone psychological.

The analogy with geometry is given by Wittgenstein himself: "It is as impossible to represent in language anything that 'contradicts logic' as it is in geometry to represent by its coordinates a figure that contradicts the laws of space" (3.032). Wittgenstein's concern here is apparently more general than geometry: the notion of what can be represented in language (what is thinkable). For Kant, however, the space that underlies geometrical construction is the space in which objects in the most general sense are given to us. Hence, also, the explanation of necessity (and apriority) which Wittgenstein gives in fact reproduces Kant's: "state of affairs that contradicts logic" (Wittgenstein) and "thing that contradicts geometry" (Kant) are literally nonsensical or unthinkable notions: for Wittgenstein, nothing is excluded

---

[32] Whether Frege's version of logic-as-doctrine is committed to essential primitiveness is a question to which there is no straightforward answer; for a discussion, see Burge (1998/2005) and Jeshion (2001).

[33] This, of course, holds quite generally for mathematics, according to Kant; mathematics "does not derive its cognition from concepts, but from their construction [...]" (Kant 1781/1787/1998: A734/B762).

by logic except straightforward nonsense; for Kant, nothing is excluded by geometry except thoughts that are literally about nothing.[34,35]

It is this Kantian strategy—Kantian in the sense of the structural deduction—that puts Wittgenstein in a position to discard the normative function of logic (in the sense of logic$_F$). An instance of this is found in (*TLP* 5.4731), which asserts that recourse to *self-evidence*, essential for Russell,[36] "can only become dispensable in logic because language itself prevents every logical mistake". Logic, rather than bearing upon thought through what we might call normative rhetoric, is woven into the space of representation, thereby giving it a unique structure. Space (and time) as conditions of representation and thinking of objects, as in Kant, are replaced by the *general propositional form*. It is common to the world and the language depicting reality; it is both the "essence of the world" and the "essence of all description" (*TLP* 5.4711). It is this form that grounds inference, for example (*TLP* 5.132). If p follows from q, I can make an inference from q to p. This can be gathered only from the two propositions, which are internally related to each other and provide the only possible justification for the inference; laws of inference—"as in the works of Frege and Russell" (*ibid.*)—are not needed for the purpose, and would be senseless.

Wittgenstein did not believe in any contentual and a priori truths. The Kantianism of the *Tractatus*—its transcendental idealism, if you like—is thus "of special, insubstantial kind".[37] The general propositional form cannot be used to derive truths that hold independently of what contingent states of affairs happen to be actualized, in

---

[34] This formulation makes use of Sullivan (1996: 199–200). The difference between my reading of the geometrical analogy and Sullivan's is that Sullivan uses an essentially weaker, more liberal, notion of geometrical rule than that which, I think, was available to Kant. Given the more liberal notion, it follows that geometry is wedded to intuitive representation in a way that involves, as Sullivan puts it, a "tightening-up" of a broader conceptual space; and accordingly, that what is conceivable—thinkable, logically possible—extends further than what is intuitable (this feature characterizes Frege's notion of *Anschauung*; cf. footnote 30 above). And it also follows that "discerning the possibilities of geometrical construction" cannot be presented as a "route to an a priori order" (Sullivan 1996: 199). But there are good reasons to think that Wittgenstein's explanation of apriority is also Kant's (or the other way round). For Kant, geometrical constructions are conditions of geometrical thought. Therefore we have the identity in explanation: "What makes logic *a priori* is the *impossibility* of illogical thought" (*TLP* 5.4731); and just so for Kant: what makes geometry a priori is the *impossibility* of contra-geometrical thought.

[35] But if Kant's explanatory strategy is the same as Wittgenstein's, how is that to be reconciled with Kant's self-proclaimed role as the official critic of dogmatic metaphysics? After all, Kant holds that thought extends further than sensible thought, because when intuition is left out, what remains is the form of thought, i.e., categories, which "think objects in general" (1781/1787/1998, A254/B309); if this were not so, there would have been no problem of transcendent metaphysics for Kant to solve in the first place. However, having connected categories with "objects in general", Kant adds immediately that the form of thought, considered apart from intuition, does not "determine a greater sphere of objects", because we are not justified in assuming that any other kind of intuition than the sensible kind is possible. Thus, insofar as the focus is on *object-related thought*, there will indeed be a limitation, *from within*, to thoughts supported by construction. And it is this feature that is crucial for the Kant-Wittgenstein connection.

[36] And also for Frege: Wittgenstein (*TLP* 6.1271).

[37] I borrow this characterization—but not its substance—from Tang (2011).

the way that, according to Kant, the content endorsed in (AS) could be proved and known to be valid independently of what appearances there are filling space and time. But the general propositional form is not a purely schematic notion either. The essence of the world and of all description is to be found in what Wittgenstein took be the sound core of Frege's and Russell's new logic: the articulation of thought by means of the function-argument model and everything that can be traced back to this idea (TLP 5.47).

Furthermore, the difference between Kant and Wittgenstein on the issue of the synthetic a priori should not blind us to the fact that that their respective treatments of geometry and logic share an important consequence. Kant's account, in its barest outline, is as follows: geometrical cognition, being founded on the construction of concepts in pure intuition, presupposes space as it pertains to sensibility, space as it is "originally represented", prior to any conceptual determination[38]; this is the single, infinite (i.e., unbounded) and subjectively given space that Kant describes in the Metaphysical Exposition of Space in the Transcendental Aesthetic (1781/1787/1998, A23–25/B37–40).[39]

This two-step dependence—of geometrical representation on construction, and of construction on the pure intuition of space—is what makes a geometrical representation a *spatial picture* in Wittgenstein's sense; a straight line drawn in imagination or on a sheet of paper,[40] for example, can represent something spatial because it itself is spatial (cf. *TLP* 2.171). And this capacity for intuitive representation is grounded in a shared spatial form, which is precisely what the Kantian originally given space is.

What holds of spatial pictures and their distinctive *pictorial form* applies to pictures in general, according to Wittgenstein: just as a spatial picture cannot depict or represent its spatial form but only displays it, so too a picture in general cannot depict its pictorial form but *displays* it, or shows it forth (*weist sie auf; TLP* 2.172). Moreover, pictorial form in the most general sense is *logical form*: it is that which any picture must have in common with what it depicts (*TLP* 2.18). Thus, what has just been said about spatial and pictorial form carries over to logical form: logical form is mirrored in propositions; they *show* it, and hence it cannot be *said* (cf. *TLP* 4.12–4.1212).

In Kant, we find a parallel conclusion regarding 'ineffability'; cognition presupposes unity of consciousness, and what makes this unity possible cannot itself be made an object of cognition:

---

[38] See Kant (1787: B150fn.); for Kant's phrase "originally represented", see Allison (1973: 175–6).

[39] The bottom-line of Kant's view is explained by J. G. Schulze, Kant's contemporary and philosophical ally, in his *Prüfung der Kantischen Critik der reinen Vernunft*: "If I should draw a line from one point to another, I must already have a space in which I can draw it. And if I am to be able to continue drawing it as long as I wish, without end, then this space must already be given to me as an unlimited one, that is, as in infinite one" (as quoted in Allison 1983: 95). For a critical discussion, see Webb (1987).

[40] Cf. Kant (1781/1787/1998, A713–4/B741–2).

[T]he mere form of outer sensible intuition, space, is not yet cognition at all; it only gives the manifold of intuition *a priori* for a possible cognition. But in order to cognize something in space, e.g., a line, I must *draw* it, and thus synthetically bring about a determinate combination of the given manifold, so that the unity of this action is at the same time the unity of consciousness (in the concept of a line), and thereby is an object (a determinate space) first cognized. (Kant 1787: B137–8).

Spatial form cannot be depicted by spatial pictures. And because there is no other way of representing it, according to Kant, there is no way of representing it at all; spatiality can only be displayed by spatial pictures, by actual geometrical diagrams. The job of geometry as a diagrammatic science is to make explicit what is implicitly involved in our *capacity to intuit*.

Wittgenstein did not believe in logic as doctrine. What he thought there is or could be, though, is a *perspicuous notation* (cf. Table 4.2). It is a *Begriffsschrift* of sorts, existing in an imperfect form in Frege's and Russell's concept-scripts (*TLP* 3.325). It is a notation which is governed by logical grammar and from which the formal properties of propositions can be recognized by mere inspection (*TLP* 6.122). This feature is what makes the notation *intuitive* (*anschaulich*) in a sense that likens it to the representations in Kant's diagrammatic-visual science of geometry.[41] The point is to *make structure perceptible* and available to observation; in one case to observation of signs in use, in the other to observation of spatial objects and operations on them. Thus, when we read the *Tractatus* as containing a version of structural deduction, we are led to acknowledge, among other things, the following "proportion", which aligns the work with Kant: perspicuous notation/logical form = geometrical diagrams/spatial form. That is, the relationship of perspicuous notation to pictorial form in the most general sense (Wittgenstein) is the same as the relation that geometrical diagrams bear to the mere form of outer sensible intuition (Kant).[42]

What, then, of *normativity*? Given structural deduction, logic as doctrine and normative rhetoric are replaced by a notation that shows forth its own ground of perspicuity. And it is this feature that makes room for normativity even within Wittgenstein's conception of logic. Now, though, normativity takes the form of a *critique of language*. Thoughts (propositions with sense) are in perfect logical order: there are no illogical thoughts (*TLP* 3.03). But there is not only *language*, the totality of *Sinnvolle Sätze*. There is also what we may call *mere speech*. This occurs

---

[41] Stenius (1989) is an illuminating discussion of Wittgenstein's claim (*TLP* 6.223) that in a mathematical sign-language, it is language itself that supplies the necessary "intuitions", which is given as the answer to the question, whether *Anschauungen* are needed in the solution of mathematical equations. Much of what Wittgenstein has to say about mathematical propositions ("equations") applies, I think, to his hypothetical perspicuous notation as well.

[42] Structural deduction gives a "logical" route from Kant to Wittgenstein. Another, more "metaphysical" route is also available, too. It goes through Schopenhauer and highlights the notion of *metaphysical subject*. Evidently, linking these two routes is a task that cannot be taken up in this paper; Appelqvist (2016) gives a Kantian reading of the *Tractarian* interrelations between metaphysical subject, logical form, limit of language and ethics. Another important topic that cannot be taken up here is the actual, historical route from Kant's structural deduction to Wittgenstein. The key link here is Heinrich Hertz and his *Bild* theory; see Patton (2009), which not only links Wittgenstein to Hertz but puts the *Bild* theory in a larger context within early analytic philosophy.

when we "have failed to give a meaning to certain signs" in our propositions (*TLP* 6.53). To curb mere speech when it occurs, and to reveal it for what it really is, is the task of philosophy, and this is effected with the help of logical symbolism (*TLP* 4.112). The apriority of logic is, indeed, to be found in the impossibility of illogical thought, and all thoughts are in logical order. However, it is not, as we might say, "evident a priori", whenever I open my mouth and say something, that I have succeeded in uttering a proposition with sense. Thus, even Wittgenstein must admit: "*In a certain sense*, we cannot make mistakes in logic" (*TLP* 5.473: italics added).[43]

References

Allison, H. E. (1973). *The Kant–Eberhard controversy*. Baltimore/London: The Johns Hopkins University Press.
Allison, H. E. (1983). *Kant's transcendental idealism: An interpretation and defence*. New Haven/London: Yale University Press.
Anderson, R. L. (2005). Neo-Kantianism and the roots of anti-psychologism. *British Journal for the History of Philosophy, 13*(2), 287–323.
Appelqvist, H. (2016). On Wittgenstein's Kantian solution of the problem of philosophy. *British Journal of the History of Philosophy, 24*(4), 697–719.
Bradley, F. H. (1883). *The principles of logic*. Oxford: Oxford University Press.
Burge, T. (1998/2005). Frege on knowing the foundation. *Mind 107*(426), 305–347. Reprinted in Burge, T. (2005). *Truth, thought, reason* (pp. 317–355). Oxford: Clarendon Press.
Burge, T. (2000/2005). Frege on apriority. In P. Boghossian & C. Peacocke (Eds.), *New essays on the a priori* (pp. 11–42). Oxford: Oxford University Press. Reprinted in Burge, T. (2005). *Truth, thought, reason* (pp. 356–389). Oxford: Clarendon Press.
Burge, T. (2005). Introduction. In T. Burge (Ed.), *Truth, thought, reason* (pp. 1–68). Oxford: Clarendon Press.
Cerbone, D. (2000). How to do things with wood: Wittgenstein, Frege and the problem of illogical thought. In A. Crary & R. J. Read (Eds.), *The new Wittgenstein* (pp. 293–314). London: Routledge.
de Jong, W. R., & Betti, A. (2010). The classical model of science: A millennia-old model of scientific rationality. *Synthese, 174*, 185–203.
Frege, G. (1879–1891/1983). Logik [zwischen 1879 und 1891]. In G. Frege, *Nachgelassene Schriften* (2nd ed., pp. 1–8). H. Hermes, F. Kambartel, & F. Kaulbach (Eds.). Hamburg: Felix Meiner Verlag.
Frege, G. (1880–1/1983). Booles rechnende Logik und die Begriffsschrift. In G. Frege, *Nachgelassene Schriften* (2nd ed., pp. 9–52). H. Hermes, F. Kambartel, & F. Kaulbach (Eds.). Hamburg: Felix Meiner Verlag.
Frege, G. (1882/1964). Über den Zweck der *Begriffsschrift*. In G. Frege, *Begriffsschrift und andere Aufsätze* (pp. 97–106). I. Angelelli (Ed.). Hildesheim: Georg Olms Verlagsbuchhandlung.
Frege, G. (1884). *Grundlagen der Arithmetik: Eine-logisch mathematische Untersuchung über den Begriff der Zahl*. Breslau: Verlag von W. Koebner.
Frege, G. (1885/1984). On the formal theories of arithmetic. In G. Frege, *Collected papers on mathematics, logic and philosophy* (pp. 112–121). B. McGuinness (Ed.). Oxford: Basil Blackwell.

---

[43] I am grateful to Gisela Bengtsson for detailed comments on an earlier version of this paper, and to Hanne Appelqvist for discussions. Research for this paper was supported by a grant from the Alfred Kordelin Foundation.

Frege, G. (1893/2013). *Grundgesetze der Arithmetik: Begriffsschriftlich abgeleitet, I. Band.* Jena: Verlag von Hermann Pohle. English edition: Frege, G. (2013). *Basic laws of arithmetic: Derived using concept-script, volumes I & II.* P. A. Ebert, & M. Rossberg, with C. Wright (Eds. & Trans.). Oxford: Oxford University Press.

Frege, G. (1897/1983). Logik. In G. Frege, *Nachgelassene Schriften* (2nd ed., pp. 137–163). H. Hermes, F. Kambartel, & F. Kaulbach (Eds.). Hamburg: Felix Meiner Verlag.

Frege, G. (1915/1983). Meine grundlegenden logischen Einsichten. In G. Frege, *Nachgelassene Schriften* (2nd ed., pp. 271–272). H. Hermes, F. Kambartel, & F. Kaulbach (Eds.). Hamburg: Felix Meiner Verlag.

Frege, G. (1918/1984). Der Gedanke: Eine Logische Untersuchung. *Beiträge zur Philosophie des deutschen Idealismus,* I (2), 58–77. English edition: Frege, G. (1984). Logical Investigations. Part I: Thoughts. In G. Frege, *Collected papers on mathematics, logic and philosophy* (pp. 351–372). B. McGuinness (Ed.). Oxford: Basil Blackwell.

Frege, G. (1919). Die Verneinung: Eine Logische Untersuchung. *Beiträge zur Philosophie des deutschen Idealismus,* I (2), 143–157. English edition: Frege, G. (1984). Logical investigations. Part II: Negation. In G. Frege, *Collected papers on mathematics, logic and philosophy* (pp. 373–389). B. McGuinness (Ed.). Oxford: Basil Blackwell.

Frege, G. (1924–1925/1983). Erkenntnisquellen der Mathematik und der mathematischen Naturwissenschaften. In G. Frege, *Nachgelassene Schriften* (2nd ed., pp. 286–294). H. Hermes, F. Kambartel, & F. Kaulbach (Eds.). Hamburg: Felix Meiner Verlag.

Frege, G. (1976). *Wissenschaftliche Briefwechsel.* G. Gabriel, H. Hermes, F. Kambartel, C. Thiel, & A. Veraart (Eds.). Hamburg: Felix Meiner Verlag.

Friedman, M. (1992a). Review of Joelle Proust: Questions of form: logic and the analytic proposition from Kant to Carnap. *Noûs, 26*(4), 532–542.

Friedman, M. (1992b). *Kant and the exact sciences.* Cambridge, MA: Harvard University Press.

Gabriel, G. (2005). Frege, Lotze, and the continental roots of analytic philosophy. In M. Beaney & E. H. Reck (Eds.), *Gottlob Frege: Critical assessments of leading philosophers, vol. I: Frege's philosophy in context* (pp. 161–175). London/New York: Routledge.

Gabriel, G. (2013). Frege and the German background to analytic philosophy. In M. Beaney (Ed.), *The Oxford handbook of the history of analytic philosophy* (pp. 280–297). Oxford: Oxford University Press.

Haaparanta, L. (1988). Analysis as the method of discovery: Some remarks on Frege and Husserl. *Synthese, 77,* 73–97.

Hyder, D. (2013). Time, norms, and structure in nineteenth-century philosophy of science. In M. Beaney (Ed.), *The Oxford handbook of the history of analytic philosophy* (pp. 250–279). Oxford: Oxford University Press.

Kant, I. (1781/1787/1998). *Critique of pure reason.* P. Guyer & A. W. Wood (Eds. & Trans.). Cambridge: Cambridge University Press.

Kant, I. (1783/1902). *Prolegomena to any future metaphysics that can qualify as science.* P. Carus (Trans.). Chicago: Open Court Publishing House.

Kant, I. (1786/2002). *Groundwork for the metaphysics of morals.* A. W. Wood (Ed. & Trans.). New Haven: Yale University Press.

Kant. I. (1800/1992). The Jäsche logic. In I. Kant, *Lectures on logic* (pp. 521–640). J. M. Young (Ed. & Trans.). Cambridge: Cambridge University Press.

Kemp, G. (1998). Propositions and reasoning in Frege and Russell. *Pacific Philosophical Quarterly, 79,* 218–235.

MacFarlane, J. (2002). Frege, Kant, and the logic in logicism. *The Philosophical Review, 111,* 25–65.

Mezzadri, D. (2015). Frege on the normativity and constitutivity of logic for thought I–II. *Philosophy Compass, 10*(9), 583–600.

Patton, L. (2009). Signs, toy models, and the a priori: From Helmholtz to Wittgenstein. *Studies in History and Philosophy of Science, 40,* 281–289.

Reck, E. H. (2007). Frege on truth, judgment, and objectivity. *Grazer Philosophische Studien, 75*, 149–173.

Ricketts, T. (1986). Objectivity and objecthood: Frege's metaphysics of judgment. In L. Haaparanta & J. Hintikka (Eds.), *Frege synthesized* (pp. 65–95). Dordrecht: D. Reidel Publishing Company.

Russell, B. (1903). *The principles of mathematics*. Cambridge: Cambridge University Press.

Searle, J. (1969). *Speech acts*. Cambridge: Cambridge University Press.

Stenius, E. (1989). *Anschauung* and formal proof: A comment on *Tractatus* 6.223. In E. Stenius, *Critical essays* (Vol. II, pp. 56–69). I. Pörn (Ed.). Helsinki: Acta Philosophica Fennica, Vol. 45.

Sullivan, P. (1996). The truth in solipsism, and Wittgenstein's rejection of the a priori. *European Journal of Philosophy, 4*, 195–219.

Tang, H. (2011). Transcendental idealism in Wittgenstein's *Tractatus*. *The Philosophical Quarterly, 61*(244), 598–607.

Taschek, W. (2008). Truth, assertion, and the horizontal: Frege on "the essence of logic". *Mind, 117*(466), 375–401.

Walsh, W. H. (1975). *Kant's criticism of metaphysics*. Edinburgh: Edinburgh University Press.

Webb, J. (1987). Immanuel Kant and the greater glory of geometry. In A. Shimony & D. Nails (Eds.), *Naturalistic epistemology* (pp. 17–70). Dordrecht: D. Reidel Publishing Company.

Windelband, W. (1912). Die Prinzipien der Logik. In A. Ruge (Ed.), *Encyclopädie der philosophischen Wissenschaften, Bd. I: Logik* (pp. 1–60). Tübingen: J.C.B. Mohr.

Wittgenstein, L. (1921/1922/1961). Logisch-philosophische Abhandlung. *Annalen der Naturphilosophie, 14*, 185–262. English editions: Wittgenstein, L. (1922). *Tractatus logico-philosophicus*. C. K. Ogden (Trans.). London: Kegan Paul, Trench, Trübner; Wittgenstein, L. (1961). *Tractatus logico-philosophicus*. D. Pears, & B. McGuinness (Trans.). London: Routledge & Kegan Paul.

Ziehen, T. (1920). *Lehrbuch der Logik auf positivistischer Grundlage mit Berücksichtigung der Geschichte der Logik*. Bonn: A. Marcus & E. Webers Verlag.

Chapter 5
Frege's Critique of Formalism

Sören Stenlund

Abstract This paper deals with Frege's early critique of formalism in the philosophy of mathematics. Frege opposes *meaningful arithmetic*, according to which arithmetical formulas express a sense and arithmetical rules are grounded in the reference of the signs, to *formal arithmetic*, exemplified in particular by J. Thomae, whose "formal standpoint", according to Frege, is that arithmetic should be understood as a manipulation of meaningless figures. However, Frege's discussion of Thomae's analogy between arithmetic and chess shows that Frege does not understand his main point, which is that we must distinguish conceptually between statements about chess figures, and the chess pieces they represent. Indeed, Thomae's fruitful use of this comparison undermines the ontological conception of arithmetic represented by Frege. The chess pieces do not go proxy for anything (as the number signs do in Frege's 'meaningful arithmetic'), but playing chess is not for that reason just manipulation of physical chess figures. This is also Wittgenstein's point when he criticizes Frege for not seeing the "justified side of formalism". Thomae's formalism can be clarified by using the distinction between sign and symbol. Symbols are never meaningless or empty forms or signs, since they have a role or function in a symbolism, which one may call their meaning. The discussion of the chess analogy shows that Frege fails to see this operative aspect of the symbolism. We can conclude that the "formal standpoint" that Frege criticizes is his fabrication, rather than anything we can attribute to Thomae.

Keywords Arithmetic • Formalism • Frege • Mathematical symbolism • Reference • Sense • Thomae • Wittgenstein

---

S. Stenlund (✉)
Uppsala University, Uppsala, Sweden
e-mail: soren.stenlund@filosofi.uu.se

© Springer International Publishing AG 2018
G. Bengtsson et al. (eds.), *New Essays on Frege*, Nordic Wittgenstein Studies 3,
https://doi.org/10.1007/978-3-319-71186-7_5

5.1

By 'formalism' I mean formalism in the philosophy of mathematics, and I shall be concerned mainly with Frege's early critique of formalism, in particular his critique of the mathematician Johannes Thomae's work. We find this criticism in sections 86–137 of the second volume of Frege's *Grundgesetze*.[1]

Thomae calls his philosophical view of arithmetic *formal arithmetic*, while Frege advocates a view he calls *meaningful arithmetic*. This 'meaningful arithmetic' is not a view of arithmetic that most mathematicians advocate; it is Frege's *philosophical* view which he articulates in his own philosophical vocabulary, where words such as 'sense', 'reference', 'object', 'function', 'thought', 'knowledge', 'judgment', 'truth-value', 'science' have their specific Fregean use. For instance, Frege (1980, § 92) says that in order to "bridge the gap between arithmetical formulas and their application", it is "necessary that formulas express a sense and that the arithmetical rules be grounded in the reference of the signs." That the signs in arithmetic have a reference to objects in an ontological realm, is precisely what Thomae rejects when he says that the signs, taken by themselves, are empty. But this emptiness of the signs is something that Frege is unable to accept, or even to take seriously.

Frege was the inventor of the quantifiers and modern formal logic, but it is important to realize that Frege, in this logical work, invented not only a formal logical system, but also a logical-philosophical vocabulary, a philosophical prose, which justifies us in talking about Frege's metaphysics.[2] Frege's notion of 'meaningful arithmetic' (which he later calls 'non-formal arithmetic') is expressed in this philosophical vocabulary, and so is his criticism of Thomae's views. The crucial point in Frege's criticism is his way of treating the following quotation from Thomae:

> The formal conception of numbers accepts more modest limitations than does the logical conception. It does not ask what numbers are and what they do, but rather what is demanded of them in arithmetic. For the formalist, arithmetic is a game with signs which are called empty. That means that they have no other content (in the calculating game) than they are assigned by their behavior with respect to certain rules of combination (rules of the game). The chess player makes similar use of his pieces; he assigns them certain properties determining their behavior in the game and the pieces are only external signs of this behavior. (Quoted in Frege 1980, § 88)

Frege is unable to understand and accept the idea that "a content is assigned to chess pieces by their behavior with respect to the rules of chess" (Frege 1980, § 95). This use of the game of chess analogy is the most central idea in Thomae's version of formal arithmetic. But Thomae is aware of the fact that the game of chess analogy has its limitations. He ends the quotation just given by saying:

> To be sure, there is an important difference between arithmetic and chess. The rules of chess are arbitrary, the system of rules for arithmetic is such that by means of simple axioms the

---

[1] I shall use the English translation by M. Black.
[2] Here I find myself in agreement with Kluge's main points in his book *The Metaphysics of Gottlob Frege* (Kluge 1980).

5 Frege's Critique of Formalism

numbers can be referred to perceptual manifolds and can thus make important contributions to our knowledge of nature. (Quoted in Frege 1980, §88)

One reason for Frege's failure to understand the idea that "a content is assigned to chess pieces by their behavior with respect to the rules of chess" seems to be that the expression 'chess piece' is ambiguous. We use it in certain contexts as 'chess figure'. A chess figure is an object of visual perception. It has a certain colour, size, shape, etc. and it is a physical object that may be made of wood. We could imagine that a child, unfamiliar with board games like chess, was familiar with the chess *figures*, even knew their names, and used them as toys in her dollhouse, but knew nothing at all about playing chess and their role in playing chess. But when we talk about their role in playing chess, how they move, or are moved, the expression 'chess piece' *has a different content*; a content which is indeed "assigned to chess pieces by their behavior with respect to the rules of chess". Frege tends to use the expression 'chess piece' only as 'chess figure', and the behavior assigned to it by the rules of chess as only external, accidental properties of the chess figures. This is very obvious in the following statement by Frege:

> For after all, chess pieces acquire no new properties simply because rules are laid down; after, as before, they can be moved in the most diverse ways, only some of these moves are in accord with the rules while others are not. (Frege 1980, § 96)

In line with this statement Frege often repeats his view that "the rules of chess treat of the *manipulation* of the pieces" (Frege 1980, § 96; italics added). Why does Frege keep using the word 'manipulation'? This word, in Frege's use of it, goes together with the notion of a chess piece as a chess figure, a visual or physical object; and the moves as being movements "in the most diverse ways" in visual or physical space.

When we say that

The bishop moves only diagonally

we are not talking about movements in visual or physical space, as if it meant: The bishop *happens to* move only diagonally, which is what Frege seems to mean when he says about the pieces that "they can be moved in the most diverse ways, only some of these moves are in accord with the rules while others are not" (Frege 1980, § 96).

This statement about the bishop is not a statement about physical movements of a figure; it is like a *grammatical statement* in Wittgenstein's sense; it is a statement that fixes the *concept* of the bishop as a chess piece. Thomae's point, when he says that "a content is assigned to the chess pieces by their behavior with respect to certain rules of combination", is that the bishop, for instance, as a chess piece is a *different concept*, with a different content from the bishop as a mere chess figure. There is a "must" involved in the rule for moving the bishop: "if it is the bishop it ***must*** be moved only diagonally", which indicates that we have to do with a concept-formation. But Frege seems to be unable to understand this, although it is the most essential point in Thomae's use of the chess analogy for explaining his formal arithmetic.

5.2

The mathematician Eduard Heine also advocated a formalist view of arithmetic. But Heine's formalism is different from Thomae's. Heine says that

> I define from the standpoint of the pure formalist and *call certain tangible signs numbers.* Thus the existence of these numbers is not in question. (Quoted in Frege 1980, § 87)

But Frege quite soon realized that this version of formalism is untenable. The sign '0' does not have the property of yielding the sign '1' when it is added to the sign '1', whatever it means to add signs, but for *the numbers* 0 an 1 we have 0 + 1=1.

Frege also raises the question what an arithmetical rule, such as a + b = b + a is supposed to mean when numbers are numerical signs. It cannot mean, for instance, that the *sign* '3 + 1' is identical with the *sign* '1 + 3'. Frege argues that sense can be made of this rule, only by stepping outside formal arithmetic, and relying upon meaningful arithmetic.

But do we have to refer to Frege's 'meaningful arithmetic', which relies upon the ontological conception of numbers as independently existing *Bedeutungen* of the number signs, to make sense of this equation? There is another aspect of formalism, which Frege does not see although it opens up a way out of Heine's view of numbers as number signs without relying upon the ontological conception of numbers. This aspect of formalism was opened up by the game of chess comparison, which was part of Thomae's formalist conception of arithmetic. Thomae's fruitful use of this comparison undermines the ontological conception of arithmetic. The chess pieces have no ontology. They do not go proxy for anything (as the number signs do in Frege's 'meaningful arithmetic'), **but still they are not dealt with as mere visual figures or wooden pieces in playing chess.**

In his conversations with the Vienna Circle Wittgenstein expressed this game aspect of formalism, and Frege's attitude to it, as follows:

> [Frege] did not see the [...] justified side of formalism, that the symbols of mathematics, although they are not signs, lack meaning [Bedeutung]. For Frege the alternative was this: either we deal with strokes of ink on paper or these strokes of ink are signs *of something* and their meaning [Bedeutung] is what they go proxy for. The game of chess itself shows that these alternatives are wrongly conceived—although it is not the wooden chessmen we are dealing with, these figures do not go proxy for anything, they have no meaning [Bedeutung] in Frege's sense. There is still a third possibility, the signs can be used the way they are in the game. (Waismann 1979, 105)

The signs *can be used* the way they are in the game, which is why they are symbols. The numbers of arithmetic and the pieces in chess are symbols, not mere signs. I shall return to the sign/symbol distinction.

5.3

So Frege could only see two alternatives in arithmetic. Either the number-signs refer to numbers as independently existing objects of an ontological realm (where they are the Bedeutungen of the signs), or else, doing arithmetic turns into the physical manipulation of signs as mere visual configurations. But by comparing arithmetic with playing chess, *Thomae recognized a third alternative*. Playing chess is not just pushing around visual figures or strokes of ink on paper. Taken by themselves the signs of arithmetic are empty, but a content is assigned to them *by their use in the arithmetical game*, by their behavior with respect to the rules of the arithmetical calculus. This is certainly the most central point in Thomae's formalism. So when Frege dismisses this point already at the beginning of his long critical account, the rest of his criticism is not very pertinent as a criticism of Thomae's views. When one reads this account one may ask oneself: Is this really a critique of Thomae's formalism?

Frege (1980, § 95) raises the question: "Can one say that a content is assigned to chess pieces by their behavior with respect to the rules of chess?" He rejects this idea as being an "inaccurate formulation", on the grounds that "neither the chess pieces nor the numerical signs have a will of their own"; "All that remains is quite simply that certain rules treat of the arithmetical figures" (Frege 1980, § 95). So the rest of Frege's criticism is based on this inaccurate understanding of Thomae's formal standpoint. Frege seems to be unable to understand how signs get a content by their use in a symbolism.

5.4

A little later in the same conversation as I quoted before, Wittgenstein says:

If we construct a figure in geometry, once again we are not dealing with lines on paper. The pencil-strokes are the same thing as the signs in arithmetic and the chessmen in chess. The essential thing is the rules that hold of those structures [...]. (Waismann 1979, 105)

In doing geometry, we are in a sense writing lines on paper. Even someone ignorant of geometry can see that, but doing geometry is something else, it is another activity. But there is another thing I want you to notice in this remark. Wittgenstein's account here comes to an end with reference to "rules that hold of those structures". This reference to rules as a sort of finishing moment in his accounts, occurs over and over again in Wittgenstein's remarks around the beginning of the 1930s. And it is often repeated by members of the Vienna Circle (e.g. by Waismann). Arguments and accounts do often end in this reference to rules with the tacit understanding that the issue in question is thereby settled. Questions about what the rules are, how they are to be understood, how they are taken to be given are not raised at this stage of Wittgenstein's thinking.

In contrast to this attitude towards rules, Wittgenstein was in the habit of questioning the significance of many established technical logical-philosophical notions by this time, such as the notions of consistency, independence, and completeness. He even questioned the Frege-Russell definition of number in terms of 'equinumerosity'. But the notion of a rule was not yet seen as in need of clarification. It may perhaps be said that there was a kind of rule-dogmatism by this time in the thinking of Wittgenstein and the members of the Vienna Circle.

How did they think of rules? The rules of chess, for instance? How were they taken to be given? As a list of rules written down in verbal language on a piece of paper?

To teach someone, who is ignorant of chess and board games, about the chess rules, you would not present him or her with a mere text. You would place a chessboard in front of him or her with the chess pieces in their initial positions, and then giving the names of the pieces, and after that you would *show* how the pieces move by actually moving them. For instance, "This is a castle, and a castle moves only straight ahead, like this."

Speaking only of rules that hold of a structure, there is always the risk of losing the *activity aspect* of chess which I think is fundamental. Chess is a game that we play, and the playing is an activity, an operative practice. And there is a chess vocabulary which is intimately connected with the activity of playing, and which the beginner must learn. It is in this vocabulary that you say, for instance, "The bishop moves only diagonally, like this". The expression 'like this' points to an actual movement of the bishop, as a *model* for how the bishop moves. This concretely exhibited model is basically what a *rule* for the movement of the bishop is like.

By saying that a chess piece moves so and so, it is not suggested that chess pieces "have a will of their own" and are agents in activities, the expression 'the chess pieces move so and so' is a special expression belonging to the game of chess vocabulary.

5.5

Let me then return to the sign/symbol distinction. It seems to me that one way of making Thomae's third alternative clearer is to use the distinction between sign and symbol that Wittgenstein uses in the first quotation I gave above (Waismann 1979, 105), and which is explained in more detail in the *Tractatus*.

There is a quite common linguistic practice among logicians and philosophers, to use the words 'sign' and 'symbol' as if they were synonymous, or at least, as if 'symbols' were a sort of signs. There is quite common to describe formalism in mathematics (the formalism that Frege criticized, for instance) by saying that

> Formalism takes mathematics to be just manipulation of meaningless (or empty) symbols.

But this use of the word 'symbol' is quite incompatible with its use in the *Tractatus*, where symbols are never meaningless. It is also in conflict with the use of the words 'symbol' and 'symbolic' that started with the work of Fransiscus Vieta. Vieta introduced the notion of 'symbol' and 'symbolic' at the beginning of the seventeenth century in connection with his invention of the algebraic symbolism.[3]

---

[3] See Vieta's "Introduction to the Analytical Art", published as an appendix to Klein (1968).

Symbols (in contrast to signs) are never meaningless or empty, since they have a role or function in a symbolism, which one may call their meaning.

The failure to see the difference between signs and symbols is most likely due to ignorance of or blindness towards the operative side of a symbolism. So here we see the importance of the activity aspect again. One source of this ignorance was the introduction of the new notion of metamathematics by Tarski, Gödel, Carnap, and others, and in particular the notion of the syntax of language. In Carnap's *The Logical Syntax of Language*, the basic signs are called 'symbols', and an *expression* is defined as a string of symbols (Carnap 1937, 4–5). Through the influence of metamathematics such a string is taken to be a *finite sequence* in the mathematical sense, so the syntax of language tends to become a branch of applied mathematics. But the finite sequences of mathematics are not in general given to us as visual objects, or objects of perception.

The syntactical form of a sign in the sense of metamathematics is no longer a visual structure, something that can be perceived, but a structure of applied mathematics which is seen as having overcome the limitations of visual structures.

In chapter 2 of Michael Resnik's book *Frege and the Philosophy of Mathematics* there is a discussion of Frege's critique of Heine and Thomae. Resnik tends to use the word 'symbol' as 'sign'. In his account of Frege's critique he renders one of Frege's objections as follows:

> Certainly, he [Thomae] cannot be asserting that the *symbol* "3 + 1" is identical with the *symbol* "1 + 3". By attributing meaning to these symbols we can make sense of this equality […]. Yet this interpretation is not open to Thomae. (Resnik 1980, 58)

What Resnik means here is that *the signs* "3 + 1" and "1 + 3" are not identical. However, these two *symbols* do indeed yield the same value, namely the number 4. And we don't need to attribute meaning to the symbols, they already have the required meaning by their use and function in the operative symbolism of arithmetic.

5.6

In the logical syntax of the *Tractatus* (Wittgenstein 1961), as well as in Vieta's use of the notion of symbol, symbols have a *non-ontological feature*. In the remark 3.317 in the *Tractatus*, Wittgenstein says:

> And *only* this is essential to the determination [of the values of a propositional variable], that it is only a description of symbols and asserts nothing about what is symbolized.

This is connected to the critique of Russell's theory of types in the *Tractatus*. In the remark 3.33 he says:

> In logical syntax the meaning [Bedeutung] of a sign ought never to play any role; it must admit of being established without mention being thereby made of the *meaning* [Bedeutung] of a sign; it ought to presuppose *only* the description of the symbols.

And in 3.331, he says:

> From this observation we get a further view—into Russell's *Theory of Types*. Russell's error is shown by the fact that in drawing up his symbolic rules he has to speak about the things his signs mean [bedeuten].

Russell's symbolic rules are therefore in a sense circular. Although, Frege did not have a theory of types, he obviously referred to the *Bedeutung* of the arithmetical signs in his understanding and justification of the arithmetical rules. So this criticism of Russell applies to Frege as well.

What then about the difference and the relationship between signs and symbols?

In the important remark 3.32, Wittgenstein says:

> The sign is what can be perceived of a symbol.

So the signs are determined as visual or perceptual structures, *but this is not true of the symbols that are expressed by the signs*. So how are symbols determined? In the remark 3.326 we find the answer:

> In order to recognize a symbol by its sign *we must consider the significant use*.

And in the next remark, 3.327, he says:

> The sign determines a logical form only together with its logical syntactical application.

These remarks appear to be a special case of the following remark (3.262):

> What signs fail to express, their application shows. What the signs conceal, their application declares.

The game of chess, as well as the calculus of arithmetic, are symbolic systems, or symbolisms. They involve the signs and their forms which are visual structures; they are a sort of things that we can perceive.

But there is another aspect of symbolisms, which is connected with the fact that playing chess as well as doing arithmetic are activities. There is an *operative side* of a symbolism, which concerns how the signs are used, how we operate with the signs in the symbolisms. It is in this operative side of the symbolism that the meaningful symbols are constituted and get their life. One might say that the meaningful symbols are the signs, not as a mere visual structures, but rather seen from the point of view of their use and function in the symbolism. Their use and function in the operative symbolism is the determining feature of a symbol. The king in chess is a symbol, and so is the number 2 in the arithmetical symbolism.

It is easy to misunderstand this operative aspect of a symbolism, since it does not have the character of being a thing or an object. It concerns, not a visual structure, but rather *how something is done*, such as how the castle is moved, or how the number 1 is added to the number 3.

It is difficult to make oneself aware of the operative aspect of a symbolism as such, without transforming it into something else; it is particularly difficult when you master the symbolism. For someone who masters, and has full control over the arithmetical symbolism, the operative aspect tends to appear as something completely trivial, which we therefore tend to be blind for.

5 Frege's Critique of Formalism

It is this difficulty that Wittgenstein is concerned with when he says in the *Blue Book*:

> The mistake we are liable to make could be expressed thus: We are looking for the use of a sign, but we look for it as though it were an object *co-existing* with the sign. (One of the reasons for this mistake is [...] that we are looking for a "thing corresponding to a substantive.") [...] As a part of the system of language, one may say, the sentence has life. But one is tempted to imagine that which gives the sentence life as something in an *occult sphere*, accompanying the sentence. (Wittgenstein 1969, 5)

Wittgenstein does not say it explicitly, but I think that he implies that the realm of references (or *Bedeutungen*) of signs in Frege's philosophy is such an occult sphere, and so is the realm of *thoughts* (which Frege called "the third realm"). These realms can be reached by a special kind of thinking and reflection, which is not ordinary mathematical activity that takes place within the symbolism. Frege's realm of *Bedeutungen* is located outside the symbolism, which is why the signs in Frege's arithmetic are not empty like Thomae's arithmetical signs. Thomae's signs, taken by themselves, are empty but they acquire content and meaning within the operational symbolism by the use and function of the symbols they are signs for.

Wittgenstein understood perfectly well Frege's problem with the mathematical formalists. In the *Blue Book* he says:

> Frege ridiculed the formalist conception of mathematics by saying that the formalists confused the unimportant thing, the sign, with the important, the meaning. Surely, one wishes to say, mathematics does not treat of dashes on a bit of paper. Frege's idea could be expressed thus: the propositions of mathematics, if they are just complexes of dashes, would be dead and utterly uninteresting, whereas they obviously have a kind of life. And the same, of course, could be said of any proposition: Without a sense, or without a thought, a proposition would be an utterly dead and trivial thing. And further it seems clear that no adding of inorganic signs can make the proposition live. And the conclusion which one draws from this is that what must be added to the dead signs in order to make a live proposition is something immaterial, with properties different from all mere signs. (Wittgenstein 1969, 4)

But rather than continuing this line of thought as Frege is doing it with his theory of *thoughts*, *Sinn* and *Bedeutung*, Wittgenstein interrupts it by saying:

> But if we had to name anything which is the life of the sign, we should have to say that it was its *use*. (Wittgenstein 1969, 4)

And its use is not something that has to be added to the signs of a symbolism from outside, but something that the signs acquire in the symbolism.

5.7

Let me summarize some of the main points of my paper.

Frege presents a critique of mathematical formalism. The main target of Frege's criticism in the *Grundgesetze* is the mathematician Johannes Thomae's *formal arithmetic*. There is no comprehensive philosophical work where Thomae presents his philosophy of mathematics. In the first two introductory sections of the second edition of his book *Elementary Theory of Analytic Functions of a Complex Variable*, he states that he has a *formal standpoint*, which he contrasts with the logical standpoint of Frege and Dedekind. But what he has to say about this formal standpoint, is essentially only what we find in the short passages that Frege quotes in his critique. I suppose that Thomae meant that his formal standpoint is being used and exemplified throughout his presentation of the theory of analytic functions of a complex variable. That, however, is an aspect of Thomae's work that Frege does not deal with.

One main feature of this formal arithmetic is that it is arithmetic *without ontology*. The numerical signs have no reference to independently existing objects of a realm outside the symbolism. The existence of such an ontological realm is essential for Frege's "meaningful arithmetic", because the rules of arithmetic are grounded in the reference of the signs. It is, according to Frege, by reflection on the reference of the numerical signs that the signs acquire their meaning and that the arithmetical rules are justified. When the ontological realm is taken away, the meaning of signs is lost, and what is then left of the numerical signs, in Frege's view, are only figures and the manipulation of figures as visual or physical objects. Now, *this manipulation of meaningless figures is what Frege erroneously calls Thomae's "formal standpoint"*.

What Frege calls Thomae's "formal standpoint" is Frege's own fabrication. It is *not* Thomae's standpoint. When the ontological realm is taken away from arithmetic what we are left with, *according to Thomae*, are *not* just visual figures and the physical manipulation of figures, but signs and expressions whose meaning come from their role in the arithmetical game. They are assigned a content by their behavior with respect to the rules of the arithmetical game, as Thomae puts it.[4] And it seems to have been Thomae's game of chess comparison that made him realize this. The game of chess has no ontology, but playing chess is not for that reason just manipulation of physical figures.

So the doctrine that Frege attributes to Thomae from the beginning to the end of his critique in the *Grundgesetze, is not really Thomae's doctrine*. Maybe this is why Frege's long critique of formalism in the *Grundgesetze* is quite difficult to read.

5.8

But the dispute between Thomae and Frege did not end with Frege's critique in the *Grundgesetze*. A few years later Thomae published a short article called *Ferienplauderei* (translated as "Holiday Chat" or "Holiday *Causerie*" in the English translations), where he tries to refute Frege's critique.[5] Frege then published a "Reply to Mr. Thomae's Holiday *Causerie*" and somewhat later an article with the

---

[4] See the quotation in Frege (1980, § 95).

[5] Thomae (1971).

5 Frege's Critique of Formalism

title "Renewed Proof of the Impossibility of Mr. Thomae's Formal Arithmetic".[6] In the latter two articles, Frege's polemical tone is much sharper than in the *Grundgesetze*. There is one point in the last article I would like to comment on. Frege says, correctly I think, that

> it should really be quite irrelevant what the game-pieces look like, as long as those that are distinguished in the rules are in fact clearly distinct, and as long as the actions of the game remain possible. (Frege 1984, 349)

And he continues:

> In chess, for example, instead of castles, knights, bishops, queens, and kings, we could use pieces representing cannons, lancers, lieutenants, colonels, and generals, and could play with these just as well as with the traditional pieces. Similarly, it should be possible to play the computing game [of arithmetic] with figures that look quite different from the signs of arithmetic: e.g. these: [here Frege presents a list of 14 signs most of which have no use in arithmetic]. (Frege 1984, 349)

Although he does not say it, he must of course presuppose a list of instructions telling us for each of the new signs which one of the usual arithmetical signs it replaces, and also that two different signs in the old notation must be replaced by two different signs in the new notation. Given that, it seems to me that such a change of the notation in arithmetic would indeed be possible. But then Frege suddenly says:

> But that doesn't work. Why doesn't it work? Why must the game pieces agree with the signs of non-formal arithmetic?[7] Because formal arithmetic cannot do without the sense which its objects [signs?] have in non-formal arithmetic. [...] Accordingly, formal arithmetic presupposes the non-formal one; its pretentions of replacing the latter herewith falls to the ground. (Frege 1984, 349–350)

I think that Frege is seriously mistaken here. He fails to see that *the sense of the signs in formal arithmetic is their use.* ("A content is conferred upon them by their behavior with respect to rules of the calculating game", in Thomae's words. This is true about the signs in the old arithmetical notation as well as in Frege's suggested new notation.) Knowing the sense of the signs is to be able to operate correctly with them in the calculating game; it is knowing *how to do* something, and not knowing something about certain objects that *determine* their use. The use of the signs in the symbolism is the 'ultimate foundation'.

But don't we have to know the sense of the signs, in order to be able to follow the rules containing them and to operate correctly with them? Yes, because their sense is their use, and we are able to learn their use. The idea that the sense of the signs is something in an occult sphere that *determines* their use, is a mistaken idea. So it is clear that it would be possible for someone to learn the arithmetical game from scratch in the new notation suggested by Frege. That is the sense in which the signs are arbitrary. But Frege seems to deny that the signs are arbitrary.

---

[6] The English versions of these latter two articles are published in Frege (1984).

[7] In this article Frege says 'non-formal arithmetic' instead of using the expression 'meaningful arithmetic' as in the *Grundgesetze*.

Frege tends to see arithmetic as a body of truth about a subject matter, and not really as a calculating game. He calls the equation

$$a + b = b + a$$

a *theorem*, a true proposition or judgment, and not a rule we follow in calculating. This appears to be connected with Frege's claim that arithmetic is a science (in a sense with roots in Aristotle), and a science must have a subject matter. This ontological conception of mathematics was abandoned by Vieta and his followers at the beginning of the seventeenth century. In Vieta's symbolic mathematics, the notion of theorem is replaced by the notion of problem. Vieta's attitude was that "We are concerned with solving problems, not with proving theorems (as Euclid)."

References

Carnap, R. (1937). *The logical syntax of language*. A. Smeaton (Trans.). London: K. Paul Trench.
Frege, G. (1980). Frege against the formalists [translation of Grundgesetze der Arithmetik, Vol. II, §§ 86–137]. M. Black (Trans.). In P. Geach & M. Black (Eds.), *Translations from the philosophical writings of Gottlob Frege* (3rd ed., pp. 162–213). Oxford: Basil Blackwell.
Frege, G. (1984). *Collected papers on mathematics, logic, and philosophy*. B. McGuinness (Ed.). Oxford: Basil Blackwell.
Klein, J. (1968). *Greek mathematical thought and the origin of algebra*. E. Brann (Trans.). Cambridge, MA: M.I.T. Press.
Kluge, E.-H. W. (1980). *The metaphysics of Gottlob Frege: An essay in ontological reconstruction*. The Hague: Martinus Nijhoff Publishers.
Resnik, M. D. (1980). *Frege and the philosophy of mathematics*. Ithaca: Cornell University Press.
Thomae, J. (1971). Thoughtless thinkers: A holiday chat. In G. Frege (Ed.), *On the foundations of geometry and formal theories of arithmetic* (pp. 115–120). E. -H. W. Kluge (Trans.). New Haven: Yale University Press.
Waismann, F. (1979). *Wittgenstein and the Vienna circle: Conversations recorded by Friedrich Waismann*. B. McGuinness (Ed.). Oxford: Blackwell.
Wittgenstein, L. (1961). *Tractatus logico-philosophicus*. D. F. Pears & B. McGuinness (Trans.). London: Routledge.
Wittgenstein, L. (1969). *The blue and the brown books* (2nd ed.). Oxford: Basil Blackwell.

Chapter 6
Why Is Frege's Judgment Stroke Superfluous?

Martin Gustafsson

Abstract Frege's use of a judgment stroke in his conceptual notation has been a matter of controversy, at least since Wittgenstein rejected it as "logically quite meaningless" in the *Tractatus*. Recent defenders of Frege include Tyler Burge, Nicolas Smith and Wolfgang Künne, whereas critics include William Taschek and Edward Kanterian. Against the background of these defenses and criticisms, the present paper argues that Frege faces a dilemma the two horns of which are related to his early and later conceptions of asserted content respectively. On the one hand, if content is thought of as something that has propositional structure, then the judgment stroke is superfluous. On the other hand, if what is to the right of the judgment stroke is conceived as a sort of name designating a truth-value, then there is no consistent way to avoid construing the judgment stroke as a kind of predicate, and thereby fail to do justice to the act-character of judgment and assertion.

Keywords Assertion • Content • Frege • Judgment • Judgment stroke • Wittgenstein

6.1 Judgment and Fiction

Gottlob Frege thought that his so-called "judgment stroke", or "assertion-sign"—*Urteilsstrich* in the original German—was one of his central logical innovations.[1] Wittgenstein always disagreed. In the *Tractatus* (mistakenly calling '⊢' as a whole the judgment stroke, rather than the vertical alone), he notes that "Frege's 'judgment stroke' '⊢' is logically quite meaningless" (Wittgenstein 1961, 4.442). In the *Big Typescript*, he took the deflationist view that "Frege's assertion-sign is properly placed if it's supposed to indicate nothing more than the beginning of a sentence.

---

[1] "Judgment stroke" is the translation chosen for the original "Urtheilsstrich", both in the 1972 English translation of the *Begriffsschrift* (Frege 1972), and in the recent English translation of *Grundgesetze der Arithmetik* (Frege 2013). "Assertion sign" is Russell's term for the same sign, but seems to me inferior as a translation of Frege's German original (even if Geach uses it in his translations).

M. Gustafsson (✉)
Åbo Akademi University, Turku/Åbo, Finland
e-mail: martgust@abo.fi

[…] the assertion-sign serves the same purpose as the period in the previous sentence" (Wittgenstein 2005, 160e). And in the *Philosophical Investigations*, he argued, similarly, that the function of the judgment stroke "is that of the full-stop", and made the following observation:

> Frege's idea that every assertion contains an assumption, which is the thing that is asserted, really rests on the possibility found in our language of writing every statement in the form: "It is asserted that such-and-such is the case."—But "that such-and-such is the case" is *not* a sentence in our language—so far it is not a *move* in the language-game. And if I write, not "It is asserted that …", but "It is asserted: such-and-such is the case", the words "It is asserted" simply become superfluous. (Wittgenstein 1958, §22)

In this paper, I want to consider in exactly what sense Frege's judgment stroke might be "superfluous", or "logically meaningless". I shall not engage in detailed Wittgenstein exegesis, but instead discuss various ways in which one might try to construe the kind of point Wittgenstein is making. I shall also look at various attempts at defending Frege.

Let me start my discussion by quoting one of the aphorisms in the opening chapter of Søren Kierkegaard's *Either-Or*:

> It happened in a theater that fire started offstage. The clown came out to tell the audience. They thought it was a joke and applauded; he repeated it; the acclaim was even greater. This, I think, is how the world will come to an end: to general applause from wits who believe it is all a joke. (Kierkegaard 1992, 49)[2]

In discussions of the function of Frege's judgment stroke, some similar story is often invoked to make the point that whatever purpose this stroke might serve, it cannot be to force upon us the recognition of a genuine assertion. "Imagine this," says Donald Davidson:

> [T]he actor is acting a scene in which there is supposed to be a fire. […] It is his role to imitate as persuasively as he can a man who is trying to warn others of a fire. 'Fire!' he screams. And perhaps he adds, at the behest of the author, 'I mean it! Look at the smoke!' etc. And now a real fire breaks out, and the actor tries vainly to warn the real audience. 'Fire!' he screams, 'I mean it! Look at the smoke!' etc. If only he had Frege's assertion sign. (Davidson 1984, 269–70)

In a similar vein, commentators point out that a judgment stroke written in front of a sentence cannot guarantee that an assertion is in fact being made, since nothing prevents an actor from employing such a construction as part of his or her acting—and Frege himself clearly thinks that actors on stage are not making any assertions (cf., Frege 1979, 234; Frege 1984, 356). "At that rate", Elizabeth Anscombe notes, "it would be an inexcusable *faux pas* to make an actor write the assertion sign before a proposition on a blackboard in a play!" (Anscombe 1959, 113)—which it obviously is not.

Observations like these need not be presented, and are often not seen, as substantive objections to Frege. Rather, they are frequently taken to clarify his conception

---

[2] Translation amended. Kierkegaard refers to an actual event in a theater in St Petersburg on February 14, 1836, which was reported in Danish newspapers.

by identifying a function that he did *not* mean his judgment stroke to have. Frege himself would presumably reject the envisaged examples as beside the point by insisting on a sharp separation of logic from fiction, as he in fact does in a posthumously published survey of his logical doctrines:

> In the *Begriffsschrift* I use a special sign to convey assertoric force: the judgement-stroke. The languages known to me lack such a sign, and assertoric force is closely bound up with the indicative mood of the sentence that forms the main clause. Of course in fiction even such sentences are uttered without assertoric force; but logic has nothing to do with fiction. (Frege 1979, 198)

One might want to respond that the sort of worry voiced by Davidson, Anscombe and others, cannot be alleviated by such reference to an allegedly neatly demarcated domain of fictional discourse. Perhaps a nineteenth century European stage performance does constitute a pretty well demarcated special context, the conventions of which were institutionalized enough to mark a relatively clear distinction. However, cases where the line between fiction and non-fiction is hard to pinpoint are commonplace—Davidson mentions historical novels and *romans-à-clef* as such cases. Indeed, Kierkegaard's own development of the theater story indicates some of the complications. After all, Kierkegaard's point is that a predicament similar to the clown's may well arise in real life, in a situation as deadly serious as you may wish, namely, when the world is about to end due to the impossibility of making people aware of the catastrophe no matter how clearly and straightforwardly one tries to provide them with the relevant information.

Nonetheless, even if it is true that the distinction between logic and fiction is more intricate than Frege appears to assume,[3] it still does not seem right to use cases like Kierkegaard's clown as straightforward objections to Frege's employment of the judgment stroke. As I think Wittgenstein was well aware, the judgment stroke was not meant to function as a failsafe tool for conveying assertoric force in real-life cases where doubt about the nature of particular utterances may otherwise arise. Frege did not intend the judgment stroke to draw the line between fictional and non-fictional discourse. Rather, the judgment stroke is meant to do its work in a context where that line has *already* been drawn. Frege introduced it as a tool that has its use only against the already made background assumption that the discourse at issue is non-fictional.

Certainly, Frege is very explicit that the judgment stroke should mark a logical distinction between asserted and non-asserted contents or thoughts. However, he sees the relevant distinction as one made *within* the domain of non-fictional discourse. At the places where he introduces the judgment stroke, his emphasis is always on the point that we need a sign to highlight the difference between, on the one hand, asserting a content or a thought, and, on the other hand, merely grasping the thought, as we do when we deal with it as an hypothesis whose consequences

---

[3] It is arguable that Frege's demarcation of fiction is more intricate than the above discussion suggests (cf. Gisela Bengtsson's contribution to this volume). I allow myself to bypass such considerations here, since the complications do not affect the main point I am making about the role of the judgment stroke.

we might want to explore for scientific purposes or as a non-asserted component of some logically complex construction like a conditional or a disjunction (Frege 1952, 34, 1972, 111–112, 2013, 9). This contrast is made within the genus of non-fictional language use. By contrast, utterances made as part of a theater performance are taken by Frege as belonging to a different genus altogether. Rather than conceiving them as just one additional variety of non-asserted discourse, along with scientific hypotheses and components of conditionals and disjunctions, Frege thinks utterances in fiction constitute a sort of language use that the logician can leave aside as irrelevant to his investigations.

In line with this, many commentators have argued that the intended function of the judgment stroke as a marker of assertoric force is internal to the use of Frege's conceptual notation or Begriffsschrift, and that Frege's idea is not that it could serve as such a definite marker in any other, real-life context. As Nicolas Smith points out, the judgment stroke is not

> meant to provide a magical antidote to false assertion: *of course* an actor could precede a formula with a judgement-stroke on a blackboard on the stage without herself asserting the formula, just as she could utter a sentence in a sincere tone of voice without thereby asserting the sentence herself [...]. This is to say that *wide* contextual factors can still override the judgement-stroke—how could that possibly fail to be the case?—while still it being the case that within a passage of the Begriffsschrift, the judgement-stroke alone is the unambiguous marker of assertoric force. (Smith 2000, 172–173)

I think this observation is correct and important. And yet, I suspect, with Wittgenstein, that there is still a legitimate worry one might have about the alleged need or usefulness of a sign for judgment or assertion, even within Frege's work. In what remains of this paper, I shall try to clarify what this worry might be.

6.2 Judgment, Content, and the Role of the Horizontal

I have argued that cases like Kierkegaard's clown do not work very well as objections to Frege's use of the judgment stroke, if the proposed criticism proceeds from the wrongheaded idea that Frege thinks the judgment stroke has the power to draw the line between fiction and non-fiction, or between serious and non-serious discourse. According to Frege, that line is not one that is drawn *within* logic, by means of the judgment stroke. Rather, it is a line that is supposed to have already been drawn as our logical investigations begin.

But now, might there be some other way in which cases like Kierkegaard's clown cause problems for Frege? One thing that is brought to the fore in such examples is that judging and asserting are *acts* and are not themselves part of what is being judged or asserted. What Kierkegaard's clown is telling his audience is not *that he is asserting that the theater is on fire*, but *that the theater is really on fire*. His statement is not a statement about what sort of speech act he is performing. Rather, it is a straightforward description of a non-linguistic state of affairs. Might this kind of point serve as a criticism of Frege's conception of the judgment stroke?

Here the suggested criticism—whatever it may be, more precisely—appears even more obviously off target. After all, Frege himself is very clear precisely on the point just made: the activity of judging and asserting must be sharply distinguished from what is being judged or asserted. In "Function and Concept", he says: "This separation of the judging from the subject matter of judgement [Diese Trennung des Urteilens von dem, worüber geurteilt wird] seems to be indispensable," (Frege 1952, 34; translation amended) and he goes on to note that "The [judgement stroke] cannot be used to construct a functional expression; for it does not serve, in conjunction with other signs, to designate an object. '⊢ 2+3=5' does not designate anything; it asserts something" (1952, 34, footnote). In his posthumously published "Introduction to Logic" from 1906, he starts off in a similar vein, warning against confusing judging with predicating (Frege 1979, 185). It is true that in the *Begriffsschrift*, he at one point says that the judgment stroke taken together with the horizontal content stroke "is the common predicate for all judgments" (Frege 1972, 113; italics removed), but the surrounding context strongly suggests that he does not use the term "predicate" in the usual way here, and that this remark cannot be read as claiming that the judgment stroke adds anything to the content which is being judged or asserted.

Indeed, Frege's very notation is designed precisely to mark the dissociation of assertoric force from asserted content. The vertical judgment stroke is placed to the left of the content stroke (or the "horizontal" as Frege would later call it (Frege 1952, 34, 2013, 9)), and is thereby clearly separated from the ensuing specification of the content or thought:

$$\vdash A$$

Could it be made any more perspicuous that whatever mistakes Frege might have made, conflating the act of judgment or assertion with the content that is being judged or asserted is not among those?

On second thoughts, however, things might start to look a bit less clear-cut. After all, it is not a very unusual phenomenon in philosophy that a philosopher insists on a certain distinction and tries to make it as clear and sharp as possible, and yet fails to do justice to its depth. A classic example is the dualist attempt to capture the distinction between the physical and the mental by saying that they constitute two radically different substances—an attempt which leads to all sorts of well-known difficulties in connection with the concept of mental substance, precisely because that notion, as the dualist conceives of it, still retains too many of the features we need to disassociate from the concept of the mental in order to get clear about its nature. Wittgenstein's criticism of philosophical conceptions often involves this sort of point: as he puts it, philosophical problems tend to arise when we make certain differences *too slight*. As Cora Diamond has pointed out (Diamond 2010), Frege, in his criticisms of traditional conceptions of logic, sometimes also makes this sort of objection. As an example, Diamond mentions his criticism of the idea that the distinction between proper names and concept-words can be captured by saying that whereas a proper name like "Jupiter" designates a determinate object—namely, the planet Jupiter—a concept-word like "planet" also designates an object, but an object of a peculiar, *in*determinate or *variable* sort. According to Frege, this way of distinguishing between proper names and concept words is not deep enough, and thus leads to philosophical puzzlement. It is, after all, quite mysterious what an "indeterminate" or "variable" object is supposed to be. Frege's solution is to make the distinction even deeper, by making it clear that a concept-word designates no sort of object at all, but a concept.

Now, the question I want to explore is whether Frege's conception of the judgment stroke involves a failure of the sort just described. Frege certainly introduces the judgment stroke with the intention to mark an important distinction. However, might it nonetheless be the case that this attempt fails because it makes the difference between the act of assertion and the asserted content too slight?

There are different imaginable versions of such an objection to Frege's employment of the judgment stroke. One version targets specifically the function of the judgment stroke in Frege's mature works, including "Function and Concept" and the *Grundgesetze*. In these works, he had arrived at the conception of non-asserted sentential constructions as names of truth-values, of thoughts as the senses of such names, and of the horizontal stroke as a function that maps the True onto itself and everything else onto the False. The objection is that this conception makes it impossible to avoid giving the judgment stroke a function similar to that of a predicate that contributes to the content of what is asserted, despite Frege's insistence to the contrary. For, according to Frege's mature conception, the role of the judgment stroke is to transform what is in effect conceived as a (complex) name—'— 2+3=5', say—into an assertion. But how can that happen, unless the judgment stroke adds to this name something like a predicate, so that we get something that can be asserted in the first place? I mean, suppose, with Frege, that '2+3=5' is a name of the True, and, consequently, that '— 2+3=5' also designates the True. Now, consider what happens when we add the judgment stroke:

$$\vdash 2+3=5$$

According to Frege, it is only by means of this whole that an assertion is finally being made: the assertion *that 2+3=5* (Grundgesetze §5); or, as it puts it in "Function and Object", with '|—2+3=5' "we are not just writing down a truth-value […] but also at the same time saying that it is the True" (1952, 34). But then, one might object, the content of this assertion is *not* the same as what Frege associated with the complex name '— 2+3=5'. Rather, what gets asserted is something that has a propositional structure, a structure of the form: *things are thus-and-so*. And how could what is asserted have such a structure, unless the name '— 2+3=5' has been supplemented with a genuine predicate—a predicate which contributes in a substantive way to what is being asserted? And, since the judgment stroke is the only sign that has been added to the name in order to transform it into an assertion, what else could have provided us with this predicate? (Cf. Ricketts 2002, 239–240).

Tyler Burge has defended Frege against this sort of objection, by arguing that it fundamentally misunderstands the special function of the horizontal stroke within Frege's mature system (Burge 2005, 108–115). According to Burge, the role of the horizontal is not just to map the True onto itself and everything else onto the False. What is really crucial to the horizontal, says Burge, is that it *also* creates what is in effect a judgeable content. The idea is this. Informally, the horizontal means "is the True" or "is a fact", and the adding of it to a name therefore gives rise to what is in effect a full sentence. For example, whereas '2+3=5' is a mere term whose Bedeutung is the True, '— 2+3=5' is a *sentence* which says *that 2 + 3 = 5 is the True*. Similarly, whereas '2+3=6' is a mere term whose Bedeutung is the False, '— 2+3=6' is a false *sentence* which says *that 2 + 3 = 6 is the True*; '5' is a mere term whose Bedeutung is 5, whereas '— 5' is a false sentence which says *that 5 is the True*; and so on. Burge emphasizes that the judgment stroke can be added only once the horizontal is in place—constructions such as '|2+3=5' or '|5' are simply ill-formed gibberish according to the rules of Frege's notation. So, according to Burge, Frege's notation does full justice to the point that mere terms cannot be judged true or false. Once the horizontal is there, what we have is no longer a mere term, but a complete sentence; and hence, adding the judgment stroke is indeed a matter of asserting something which *already* has propositional structure. Consequently, Burge argues, the above objection is entirely off target.

One problem with Burge's interpretation is that it is hard to square with Frege's view that "the word 'true' has a sense that contributes nothing to the sense of the whole sentence in which it occurs as a predicate" (Frege 1979, 252). A more fundamental (though perhaps ultimately closely related) difficulty, however, is one identified by William Taschek—namely, that the position Burge ascribes to Frege only serves to hide the fundamental difficulty which the objection above was gesturing at. The crucial question is if the horizontal can really play the sort of special double role that Burge's Frege burdens it with. Can it really *both* signify a (first-level) function which maps the True onto itself and everything else on the False, *and* be the source of genuine propositional structure? As Taschek points out, the difficulty with ascribing to the horizontal such a special double role is indicated by the fact that this gives rise to anomalies of substitution in Frege's notation. In a logically adequate notation, a first-level functor should of course always be replaceable by another first-level functor. However, since other first-level functors do not have the ability to create propositional structure in the way the horizontal is supposed to have, anomalies immediately arise. A simple and straightforward example illustrates the point: if we replace the horizontal in the well-formed formula '|— Φ(a)' with the first-level concept expression 'Φ', we get the ill-formed '|Φ(Φ(a))' (Taschek 2008, 395–396).

And this is not a problem which can be fixed by some additional technical adjustment. It is a reflection of a basic logical point: the structure of a complex name and the structure of a proposition are fundamentally different. As Taschek puts it,

> having obliterated the distinction, Frege—with a touch of technical genius—uses his horizontal to try, in effect to recapture it. But [...] this only works if the horizontal itself is viewed as having an anomalous (not merely distinctive) logical status among first-level concept expressions. [...] Once the demand is made that only expressions prefaced by the

horizontal can be judged or asserted—or for that matter, negated or conditionalized, or quantified into—the horizontal has, in effect, been assigned a logical role that distinguishes it sharply from all other first-level functors. (Taschek 2008, 397)

Taschek's emphasis on a fundamental logical distinction between propositional and sub-propositional structure seems to me very much in line with early Wittgenstein's worries about Frege's use of the judgment stroke.

Remember, however, that this sort of objection is targeted specifically at Frege's use of the judgment stroke in his mature works. It does not work against the conception presented in the *Begriffsschrift*, where Frege had not yet assimilated sentences to names. In the *Begriffsschrift*, Frege allows the judgment stroke to be prefixed only to what he calls "judgeable contents" ("beurtheilbare Inhalte"), and such contents are given by sentences *rather than* names (Frege 1972, 112).[4]

However, one might still worry that already in the *Begriffsschrift*, there is a problem about Frege's introduction of the judgment stroke—a more fundamental and general problem that arises because the very idea of using a *sign* to make the difference between the act of assertion and asserted content means that this difference is made too slight. In his recent book on Frege, Edward Kanterian criticizes Frege's judgment stroke along just such lines. According to Kanterian, Frege is right that judgment and assertion are acts, and that it is important to distinguish them from what is being judged or asserted.[5] However, when Frege tries to mark that distinction by means of a sign, Kanterian argues, he obscures the real significance of the difference that is at stake. According to Kanterian, "the assertion consists in utter*ing* or writ*ing* down the sentence, that is, what is being asserted, but these acts of assertion do not add any sign to what is uttered or written" (Kanterian 2012, 59). He goes on to note that even in a context where prefixing an assertion sign to sentences is common practice, it would still be my *writing down*, say, '|— It is raining', that effected the assertion—not the assertion sign itself. As Kanterian puts it, "[a] sign cannot perform its own or some other sign's assertion, since it cannot perform anything at all" (Kanterian 2012, 60; notes omitted).

In order to spell out in greater detail the worry I think Kanterian is articulating, it will be useful to look at how an interpreter who is more sympathetic to Frege's judgment stroke tries to handle the point that assertion is an act and thus not part of

---

[4] Taschek, however, very elegantly and convincingly traces the tension within Frege's mature system to a tension which is arguably present in his earlier works too, between a universalist-descriptive and a normative conception of logic. Making this connection also strikes me as a characteristically Tractarian move, but I cannot pursue this issue here.

[5] One might think that both Kanterian and Nicholas Smith (discussed below) over-emphasize the act-character of Fregean judgment. Surely, assertion is an act, which we can choose to do or not to do, and which is intentional. By contrast, judging is merely a matter of holding-true, and usually we do not chose or intend to hold specific propositions true. However, the important thing for both Kanterian and Smith seems to be the availability of the distinctions between judging and what is judged, and between asserting and what is asserted—which seems indisputable. Moreover, central normative notions apply to both judgment and assertion: they are essentially goal-directed acts aiming at truth, which in turn defines obligations of coherence, withdrawal in the light of counter-evidence, and so forth. Cf. Kremer (2000, 580).

what is asserted. Just like Kanterian, Nicolas Smith puts great emphasis on Frege's notion that judgment and assertion are acts. However, Smith thinks there is nonetheless a genuine need for the judgment stroke, due to the fact that Frege's Begriffsschrift is a *written* symbolism: "[Frege's] system comprises marks on paper. Hence the need to encapsulate *the act of putting a sentence forward* in what looks like just another symbol" (Smith 2009, 643; original italics). To make the alleged notational necessity vivid, Smith asks us to

> imagine a movie version of *Begriffsschrift* or *Grundgesetze*. There are things written on cards, which are seen propped on an easel long enough for us to read each one. Then at some points Frege picks up a card and thrusts it towards the camera—or taps it with a pointer while giving it a meaningful look. In this movie version, we never see the judgement stroke as a written symbol on any of the cards. Rather, its role is played by Frege's actions. Now in a written symbolism we do not have available such actions, so everything has to be written as a symbol. But it is crucial that the judgement stroke be thought of not as a symbol alongside the symbols for negation, the conditional, and so on, but as embodying or representing an *action*. So wherever you see the judgement stroke, think of it not as just another component of a sentence which sits before you on the page: think instead of the sentence—beginning *after* the judgement stroke—as highlighted, or as jumping out of the page at you. If Frege had written in HTML—the language of web pages—he might well have used flashing text in place of the judgement stroke! (Smith 2009, 643)

I have quoted this passage at length, because I find it quite intriguing. What, exactly, is Smith after here? Of course he is aware that it would have been a nightmare to actually read the text of the *Begriffsschrift* had Frege used flashing light instead of the judgment stroke. So what, exactly, does Smith suppose would have been gained by such a convention? For it is, after all, just another convention—as it would be just another convention to say that throwing a card towards a camera, or tapping it with a pointer while giving it a meaningful look, amounts to making an assertion. Perhaps Smith is right that such conventions would make it easier to keep in mind that what Frege is trying to mark is a kind of action rather than a piece of content. However, that seems to be just an observation about the psychological or pedagogical suitability of different notations. Logically speaking, all these different conventions seem on a par: none of them is any better than the other at actually *effecting* an assertion.

Perhaps Smith would agree with this point. Presumably, what he wants to say is that Frege thinks of judging and asserting as acts, and that, therefore, if we want a notation which marks such acts, it is useful to have one which is as different as possible from the sort of notation we use to capture the content that is being asserted. However, there is still the peculiar idea that the fact that Frege is working with a *written* symbolism somehow requires that he make use of a conventional sign of assertion. *Prima facie*, it is not at all clear why there is any such need: works of non-fictional prose, including works of mathematics, are full of assertions successfully made and communicated without the use of any sign for judgment or assertion. The point that assertion is an act does not entail that hearers and readers must be present when the act actually performed in order to grasp its assertoric force without further ado. Writing a piece of non-fiction means making a lot of assertions, assertions that might be read and understood as such years and years later, even if the written sentences are not prefixed with any sign for judgment or assertion. The sort of principled distinction between oral and written communication of assertoric force that Smith seems to be envisaging is unfounded.

6.3 A Dilemma for Frege

There is, however, more to say in defense of the view that Frege needs the judgment stroke. Arguably, my discussion so far has taken too little account of Frege's own, and by the lights of contemporary standards rather foreign, conception of the nature and aims of logic. Let us now look at one more attempt to make sense of his judgment stroke—an attempt that connects Frege's conception with his overall view of logic. Wolfgang Künne has argued that in order to understand why Frege needs his judgment stroke, we must remember two things. First, we need to keep in mind that for Frege, it is essential to logic that it is a pursuit of *truth*, and, hence, that the study of logical inference is the study of what can be proved from *true* premises that are judged to be true. Second, we must remember the demand for full explicitness in Frege: Frege takes it as crucial to a proper Begriffsschrift that its notation puts on display every logically significant element. According to Künne, these two points explain the significance of the judgment stroke for Frege:

> Why is it good to have such a sign in one's language? The raison d'être for an ideography is to serve as a medium for writing down gapless *proofs*. If one proves something one's premises are truths that one takes to be truths. Something that is essential to a proof would not be represented if the representation of a proof in an ideography would not mark premises and conclusions as put forward as true. (Künne 2009, 55–56)

This is sensitive to crucial elements in Frege's thought, and seems to offer a plausible response to Kanterian's worry, that "[a] sign cannot perform its own or some other sign's assertion, since it cannot perform anything at all". If Künne is right, the point of the judgment stroke is not to *perform* or *effect* an assertion in any philosophically deep sense, but merely to *mark* or *make explicit* assertoric force in a conventional fashion within a constructed notation. And it seems hard to deny that we can make assertoric force explicit in such a way if we want to, just as we can make interrogative force explicit by means of a question mark, for example.

This is a fairly innocuous interpretation of Frege's basic reason for introducing the judgment stroke; and it seems to me quite plausible. And yet I think we can still raise a question about whether we have here been given a reason to think that Frege really *needs* the judgment stroke in order to make assertoric force explicit.

What I mean is this: when we talk about marking or making explicit assertoric force, what we mean is to mark or make explicit the *distinction* between asserted and non-asserted judgeable content. However, to make a distinction explicit, it suffices to mark *one* of the two things that are being distinguished. It follows that even if we agree with Frege that we should have a notation that marks explicitly the distinction between asserted and non-asserted content, we could instead introduce a special sign for the latter—that is, a sign that marks a content's *lacking* assertoric force. Indeed, something like this seems to be what we have in ordinary language: if we put forward a sentence without asserting it, we prefix it with constructions such as "let's assume", "suppose that", "let us hypothesize", and so on. And in cases where the non-asserted content is part of a disjunction or a conditional, the sign for disjunction or for implication by itself suffices to mark it as without assertoric force.

In "Function and Concept", Frege writes that the reason why the separation of the act from the subject matter of judgment is indispensable is that "otherwise we could not express a mere supposition—the putting of a case without a simultaneous judgement as to its arising or not" (Frege 1952, 34). And then he immediately concludes: "We thus need a special sign in order to be able to assert something" (ibid.). What I am suggesting is that the conclusion does not follow, if the special sign is presumed to be a sign marking assertoric force. If our aim is just to mark the distinction Frege is talking about, a sign marking the lack of assertoric force would do just as well.

In fact, perhaps a case can even be made for saying that such an alternative notation is preferable to Frege's judgment stroke. Remember again the earlier criticism of how the judgment stroke was meant to function in Frege's mature works. This criticism was to the effect that if what is to the right of the judgment stroke is thought of as a *name*—as it is in Frege's mature works, namely, the name of a truth value—then we will have insurmountable difficulties upholding the crucial distinction between the structure of a complex name, and the structure of a genuine proposition. Like Frege in the *Grundgesetze*, we will have to hover between these two notions of complexity in order not to turn the judgment stroke into what is in effect a predicate, and thereby obfuscate the act-character of judgment. On the other hand, if we consistently think of that which is to the right of the judgment stroke as something which has propositional structure *rather* than the structure of a complex name—as Frege does in the *Begriffsschrift*—then it seems we have something which will already function as an assertion once it is put forward, even if there is no judgment stroke in front of it.

My suggestion, then, is that Frege is facing a dilemma. *Either* he construes what is to the right of the judgment stroke along the lines of his mature works—that is, as a name of a truth-value. Then, in order to take into due account the act-character of judgment, he will be under virtually irresistible pressure to obliterate the distinction between propositional structure and the sort of structure involved in a complex name. *Or*, he consistently construes what is to the right of the judgment stroke as something with propositional structure *rather than* as a name. But then, what is to the right of the judgment stroke is something to which a judgment stroke does not need to be added for a judgment or assertion to be made; for, in the default case, the mere putting forward of the sentence itself will suffice. Typically it is only if we are putting it forward *without* making an assertion that we need to mark what we are doing by adding something in front of the sentence, a reservation like "suppose that". If this is right, Wittgenstein was correct to think that the judgment stroke is both logically superfluous and philosophically misleading.

The only remaining reason for why one might want to make use of a judgment stroke seems to be purely notational. And this, I take it, is essentially Wittgenstein's view: the judgment stroke is really nothing more than a mark of punctuation of a certain sort—something like our full stop, only that it is placed at the beginning of sentences rather than at the end. Our ordinary full stop serves to organize written language so that we know where one sentence finishes and the next begins. Given the peculiar layout of Frege's notation, a vertical stroke at the beginning of formulas serves this sort of organizational purpose much better than a full stop at the end. Indeed, the judgment stroke is a key to the perspicuity and readability of a page full of Begriffsschrift formulas. It seems clear, however, that to offer this as the raison-d'être for the judgment stroke would be to concede that this sign fulfills no essential philosophical or conceptual purpose, but is only a notational tool whose importance is a product of the particular notational conventions that Frege happens to follow. Again, this is probably Wittgenstein's assessment; but it seems clear that Frege himself would not be willing to take such a deflationary view.[6]

References

Anscombe, G. E. M. (1959). *An introduction to Wittgenstein's Tractatus*. London: Hutchinson & Co..
Burge, T. (2005). *Truth, thought, reason: Essays on Frege*. Oxford: Clarendon Press.
Davidson, D. (1984). *Inquiries into truth and interpretation*. Oxford: Clarendon Press.
Diamond, C. (2010). Inheriting from Frege: The work of reception, as Wittgenstein did it. In T. Ricketts & M. Potter (Eds.), *The Cambridge companion to Frege* (pp. 550–601). Cambridge: Cambridge University Press.
Frege, G. (1952). Function and concept. In M. Black & P. Geach (Eds.), *Translations from the philosophical writings of Gottlob Frege* (pp. 21–41). Oxford: Basil Blackwell.
Frege, G. (1972). *Conceptual notation and related articles*. T. W. Bynum (Trans.). Oxford: Clarendon Press.
Frege, G. (1979). *Posthumous writings*. P. Long & R. White (Eds.). Chicago: Chicago University Press.
Frege, G. (1984). *Collected papers on mathematics, logic, and philosophy*. B. McGuinness (Ed.). Oxford: Blackwell.
Frege, G. (2013). *Basic laws of arithmetic*. P. A. Ebert, & M. Rossberg (Trans.). Oxford: Oxford University Press.
Kanterian, E. (2012). *Frege*. London: Continuum.
Kierkegaard, S. (1992). *Either/or. A fragment of life*. A. Hannay (Trans.). London: Penguin Books.
Kremer, M. (2000). Judgment and truth in Frege. *Journal of the History of Philosophy, 38*, 549–581.

---

[6] Thanks to Gisela Bengtsson and Silver Bronzo for very valuable comments. Earlier versions of this paper were presented at the University of Bergen, Åbo Akademi University, and the University of Chicago. Thanks to the audiences for their comments. Work on this paper was financed by the Academy of Finland, project #267141.

Künne, W. (2009). Wittgenstein and Frege's logical investigations. In H. J. Glock & J. Hyman (Eds.), *Wittgenstein and analytic philosophy: Essays for P. M. S. Hacker* (pp. 26–62). Oxford: Oxford University Press.

Ricketts, T. (2002). Wittgenstein against Frege and Russell. In E. Reck (Ed.), *From Frege to Wittgenstein* (pp. 227–251). Oxford: Oxford University Press.

Smith, N. J. J. (2000). Frege's judgement stroke. *Australasian Journal of Philosophy, 78,* 153–175.

Smith, N. J. J. (2009). Frege's judgement stroke and the conception of logic as the study of inference not consequence. *Philosophy Compass, 4,* 639–665.

Taschek, W. W. (2008). Truth, assertion, and the horizontal: Frege on "the essence of logic". *Mind, 117,* 375–401.

Wittgenstein, L. (1958). *Philosophical investigations* (2nd ed.). G. E. M. Anscombe (Trans.). Oxford: Basil Blackwell.

Wittgenstein, L. (1961). *Tractatus Logico-Philosophicus*. D. F. Pears, & B. McGuinness (Trans.). London: Routledge.

Wittgenstein, L. (2005). *The big typescript: TS 213*. C. E. Luckhardt, & M. A. E. Aue (Eds. & Trans.). Oxford: Blackwell.

Chapter 7
Frege on Dichtung and Elucidation

Gisela Bengtsson

Abstract In this paper, I identify an assumption at play in anti-semantic interpretative approaches to Frege: the notion that translatability to Frege's concept script functions as a criterion for deciding whether a thought is expressed in a sentence or utterance. I question the viability of this assumption by pointing to Frege's accounts of the aim and character of his logical language and scientific discourse more generally, and by looking at his remarks on poetic forms of language, literature and fiction (*Dichtung*). Since it seems clear that the sentences used in poetic and literary forms of language that Frege discusses, have *Sinn* and are possible to understand, in his view, I argue that the translatability criterion for thoughts is flawed. A discussion of Frege's appeal to an approach of willingness to understand in a reader, and the relation between Frege's use of elucidatory discourse and his conception of *Dichtung* is central to my exposition in this paper.

Keywords *Begriffsschrift* • Elucidation • Frege • Logic • "Meeting of minds" • Poetry • Rationality • Science • Thought • Understanding

7.1 Different Conceptions of Philosophy

Frege argued that logic is the most general science and worked with a conception of logic as universal. These aspects of Frege's work form a background for the common view that he gives science an all-embracing status that shapes his conception of rationality. As Frege is one of the originators of analytic philosophy, it is further natural to think of his work as representative of this philosophical approach. But there is no simple answer to the question of what characterizes analytic philosophy; a common conception is that analytic philosophers work with an understanding of philosophy as very similar to, or continuous with, science. Pivotal to analytic philosophy, when understood in this manner, is the notion that philosophical investigations are truth-seeking enterprises directed at the accumulation of knowledge

G. Bengtsson (✉)
Uppsala University, Uppsala, Sweden
e-mail: gisela.bengtsson@filosofi.uu.se

that can be expressed in theories.[1] Given this perspective, it's the content of the thoughts that are expressed in a philosophical text that is of interest to us, and a maximally transparent and direct form of expression of the *content* that is to be communicated stands out as vital. If we take it that philosophy and science have a common ground in their goals as well as in their methods, a form of representation that approximates what Frege refers to as a "scientific exposition" in a passage from the essay "Thought" presents itself as an ideal for philosophical investigations:

> What are called the humanities [*Geisteswissenschaft*] are closer to *Dichtung*, and therefore less scientific, than the exact sciences [*die strengen Wissenschaften*], which are drier in proportion to being more exact; for exact science is directed toward truth and truth alone. Therefore all constituents of sentences not covered by the assertoric force do not belong to scientific exposition; but are sometimes hard to avoid, even for one who sees the danger connected with them. Where the main thing is to approach by way of intimation [*auf dem Wege der Ahnung zu nähern*] what cannot be conceptually grasped [*gedanklich unfassbaren*], these constituents are fully justified. The more rigorously scientific an exposition is, the less the nationality of its author will be discernible and the easier it will be to translate. On the other hand, the constituents of language to which I here want to call attention make the translation of poetry [*Dichtungen*] very difficult, indeed make perfect translation almost always impossible, for it is just in what largely makes the poetic value that languages most differ. (Frege 1918/1997, 63/330; modified translation)

A very different approach to philosophy is at work when similarities between philosophy and literature—*Dichtung*—are emphasized. The view that the content of a philosophical text can be separated from its original form to be communicated without significant loss of meaning in other texts (such as paraphrase, interpretations or summaries of main theses) is then abandoned. Abandoned is also the thought that philosophical activity should consist in debates on opposing theses and theories. Philosophers that come to mind as relevant examples here are Plato, Nietzsche, Kierkegaard, Heidegger, and Wittgenstein: It is fair to say that their texts in many respects resist translation to other forms than they have, and that they in certain respects resist translation to other languages. The relation between these philosophers' texts and their interpreters' texts can perhaps be compared to the one that holds between a poem and an interpretation of the poem. From the viewpoint of such a conception of philosophy, the irreplaceability of a text constitutes an important point of similarity between philosophy and *Dichtung*. The thought that prose which is free of literary devices and figurative forms of language, represents an ideal form of discourse, is foreign to those who work with an understanding of philosophy as closely linked to *Dichtung*. The worry is that seeking clarity in quasi-scientific prose will lead to a simplification of what is at stake in philosophical investigations and fail to clarify the many aspects that are involved when philosophical difficulties are confronted.

I have now spoken of a conception of philosophy as closely tied to science, as opposed to a view of philosophy as closely related to *Dichtung*. But we could perhaps also say, with reference to some of Gottfried Gabriel's discussions, that it is a

---

[1] Cf. here e.g. von Wright (1994), Quine (1981), and Glock (2008).

question of different kinds of reasoning[2]: In the first case the paradigm is reasoning by way of logical inferences and proofs, in the other case the reasoning proceeds mainly by way of analogies, metaphors and similes. In the first form of reasoning, we seek to uncover logical differences and discern logically different parts of thoughts and their relations by way of analysis of propositions and arguments. The second form of reasoning can be described as analogical, in that our focus is directed at similarities between cases rather than at differences, through the literary forms of language that are used. Part of my aim in this paper is to raise the question as to where we find Frege's philosophical work, with regard to the two conceptions of philosophy that I have outlined. I will approach this question by looking at different forms of discourse that Frege discusses and uses: (1) Scientific forms of discourse, and connected with this, the logical language that he presents, (2) elucidations in the form of sentences and prose [*Erläuterungen*], (3) communication in the language of everyday life [*Sprache des Lebens*][3] and (4) *Dichtung*.

7.2 A Purely Logical Language

In the *Preface* to *Begriffsschrift*, Frege makes it clear that his logical language is put forth as a tool [*Hilfsmittel*] to test the validity of chains of inference, and clearly bring out every presupposition that has been made to secure gap-free proofs (Frege 1879, V). The primary area of use for this language is within the science(s) of logic and arithmetic, but he thinks that the area of use can be extended to geometry and possibly other sciences such as pure theory of motion, mechanics and physics. In a famous passage, he suggests that the *Begriffsschrift* might, in a developed form, also become useful for philosophers. He writes:

> If it is a task of philosophy to break the domination of words [*Herrschaft des Wortes*] over the human spirit, by laying bare illusions that through the use of language often almost unavoidably arise concerning the relations of concepts, and by freeing thought from that with which only the means of expression of ordinary language, constituted as they are, saddle it, then my *Begriffsschrift*, further developed for these purposes, can become a useful tool for philosophers. Admittedly, as is surely inevitable in the case of external means of representation [*bei einem äussern Darstellungsmittel*], even this [language] cannot make thought pure again; but the deviations can, at least, be limited to the unavoidable and harmless, whilst at the same time, just because they are of a quite different kind from those typical of ordinary language, protection is provided against the one-sided influence of one of these means of expression. (Frege 1879, VI–VII; modified translation, cf. Beaney's translation in Frege 1997, 48–49, and Bauer-Mengelberg's translation in van Heijenoort 1967a, 7.)[4]

Even though Frege points out here that the logical language is a form of representation which is external to thought (just as other languages), and that this

---

[2] Gabriel and Polimenov (2012).
[3] Frege (1879, V).
[4] Cf. here Wittgenstein (2001, §§ 90, 110, 593).

language "cannot make thought pure again", the quoted passage illustrates the centrality of he idea of a maximally clear and transparent form of representation of thoughts to his project. It touches on some of Frege's objections to the use of natural language in scientific investigations: a natural language, such as German, is not in agreement with the true nature of thought (*das eigentliche Wesen der Denkens*) and its linguistic means of expression give rise to illusions. The logical language shall serve to make features visible that are hidden or distorted when we follow the forms of expressions of natural language too far or too closely in our work in logic or mathematics. The incongruence between the true nature of thought and the grammar of natural languages is, according to Frege, a result of the cooperation of several dispositions in man when language was created. Among these dispositions are our logical disposition and our poetic disposition. If man had followed only the former when creating language, there would perhaps be no need for logic (as a science), Frege speculates (1969/1979, 154–155, 289/142–143, 270). In a late essay he suggests that if our language had been logically more perfect, we might have been able to "read [logic] off from the language".[5] His dissatisfaction with natural language also concerns its imprecise terminology and our fluctuating use of words. The latter has the effect that there is no regular connection between the *Sinn* and the *Bedeutung* of a word in the language of life.[6] A regular connection between the *Sinn* and the *Bedeutung* is however required in scientific terminology, according to Frege, and mutual understanding among investigators about the meaning of primitive terms of the science they work with, is key to such regularity.[7]

The use of a terminology that pertains to a specific science among scientists in their reasoning can be described as the use of a scientific form of discourse in an activity which is directed at reaching the truth. Or, we can say that scientists who have a mutual understanding of the meaning of the terms that form the scientific terminology engage in scientific reasoning.[8] Now, the distinction Frege makes between scientific forms of discourse and other ways of using language is made with reference to the purposes that are pursued in different contexts. In science we want to reach the truth and we are only interested in statements that can form part of the system of science, according to Frege. We reach such statements in the following manner, according to his exposition in the essay "Thought": we grasp a thought and then acknowledge the truth (or falsity) of the thought in an act of judgment. The judgment is manifested in an assertion. To the grasping of a thought corresponds a

---

[5] "If our language were logically more perfect, we would perhaps have no further need of logic, or we might read it off from the language. But we are far from being in such a position. Work in logic just is, to a large extent, a struggle with the logical defects of language, and yet language remains for us an indispensable tool. Only after our logical work has been completed shall we possess a more perfect instrument" (Frege 1969/1979, 272/ 252).

[6] Cf. Frege (1892a/1997, 27–29/153).

[7] "Since mutual cooperation in a science is impossible without mutual understanding of the investigators, we must have confidence that such an understanding can be reached through elucidations [*Erläuterungen*], although theoretically, the contrary is not excluded" (Frege 1906/1984, 301/301).

[8] Cf. Weiner's discussion in "Frege's Unmetaphysical Story about Natural Language and Truth" (Weiner, forthcoming).

propositional question that can be answered by "yes" or "no" and the answer is expressed in an assertion (Frege 1918/1997, 62/329). In another famous passage, Frege gives the following account:

> An advance in science usually takes place in this way: first a thought is grasped and thus may be expressed in a propositional [yes-no] question; after appropriate investigations, this thought is finally recognized to be true. (Frege 1918/1997, 62–63/330)

Since the *Begriffsschrift* displays the laws of the most general science—logic— and Frege speaks of these laws as "the laws of thought" that we must abide by if we want to reach truth, we can see that the relationship between logic, science and reason (or rationality) is very tight. We have here a triad of nearly interchangeable concepts.[9] So, if Frege's *Begriffschrift* is a language which is in agreement with the true nature of thought, it seems that it may provide us with a kind of test: If a sentence in natural language resists translation to the concept-script, this can be taken to show that it is not a sentence that we can understand (even if we may be led to believe otherwise due to misleading features of natural language). If a sentence is not possible to understand it seems legitimate to say that it is nonsensical, or that it does not represent a case of rational discourse (when "rational" and "scientific" are used as nearly synonymous concepts, and my suggestion is that this is often the case in Frege's writings).[10]

The picture I have now roughly sketched of Frege's conception of 'rational discourse' can be drawn from his characterization of the purely logical language, his conceptions of 'thought', and of 'logic' and his critique of natural language as inappropriate for scientific activities.[11] It is a picture that can be found within the anti-semantical interpretative tradition and in the debate on the relation between Frege and the early Wittgenstein and their respective conceptions of 'elucidation'. If we turn to James Conant's writings, for instance, we find a strong emphasis on how resistance to translation to the *Begriffsschrift* reveals that no thought has been expressed in a sentence. He writes that "if we appreciate the logically fundamental character of the distinctions upon which Frege's *Begriffsschrift* is based then we will see that anything which can be thought can be expressed in the *Begriffschrift*" (Conant 2000, 181). He goes on to say that "[i]n grasping the distinction between that which can and that which cannot be expressed in *Begriffsschrift*, we furnish ourselves with a logically precise articulation of the distinction between that which

---

[9] See Frege (1884, §§ 26–27) where he ends a discussion about objectivity with these words: "Der Grund der Objectivität kann ja nicht in dem Sinneseindrucke liegen, der als Affection unserer Seele ganz subjectiv ist, sondern, soweit ich sehe, nur in der Vernunft. Es wäre wunderbar, wenn die aller exacteste Wissenschaft sich auf die noch zu unsicher tastende Psychologie stützen sollte." See also Frege (1884, § 105), especially the second footnote to this section, and Frege (1969/1979, 133/122).

[10] Cf. Frege's discussions of the relation between reason (*Vernunft*) and science in Frege (1882, 55), and Frege (1918, 74) where he suggests that "[d]em Fassen der Gedanken muß ein besonderes geistiges Vermögen, die Denkkraft, entsprechen" and goes on to discuss the relation between thinking, truth and the foundation of science. See also Frege (1969/1979, 288/269).

[11] The three principles introduced in *The Foundations of Arithmetic* should perhaps be added to this list (Frege 1884, X).

('in a strictly logical sense') is, and that which is not, a thought" (Conant 2000, 181–182 and n. 42). If we take it that Frege's purely logical language can be used as a test for whether or not we are faced with a case of rational discourse, as outlined above, we end up ascribing a rather narrow conception of 'rationality' and 'rational discourse' to Frege. One of my aims in this paper is to sketch a more complex picture of Frege's conceptions of 'rationality' and 'understanding'. Pursuing this aim will shed light on some the questions that I raised in the introductory section of this paper about the status of Frege's conception of 'philosophy'. For these purposes we will now look at Frege's use of elucidatory sentences and prose.

7.3 Elucidation

If we consider that Frege gives the most general science—logic—the most fundamental role in our search for knowledge,[12] we might assume that this will be reflected in the form of his discursive writings. That is, we might expect him to primarily use a form of discourse that is liberated from stylistic ornaments and figurative modes, i.e. elements that have no place in a scientific exposition. But when we study Frege's writings, we see that the use of metaphors, analogies and figurative modes of speech is frequent.[13] The plentiful use of figurative modes of speech is connected with some main aspects of Frege's work: his conception of logic leaves no recourse to an intuitive logic that allows us to engage in theorizing *about* the universal framework logic forms as from a standpoint external to it.[14] The dilemma that arises was discussed as "the logocentric predicament" in Heijenoort's paper from 1967, and it has been further discussed and clarified by for instance Ricketts (1985), Weiner (1990), Diamond (1991), Goldfarb (1997), and Floyd (1998). To shed light on Frege's view of logic, Diamond (1991, 29–30) has suggested that Frege's purely logical language presents displays *the Mind*, i.e. our understanding when it is in agreement with itself and not something external to it.[15] Another way of describing what the *Begriffsschrift* displays, suggested by Alnes, is to say that it makes the basic laws of reasoning transparent to us, i.e. the laws of thought that "cover everything that can be correctly and meaningfully said, or even thought" (Alnes 1998, 119).[16]

Communicating the purely logical language to others is connected with difficulties: The clarification of the meaning of the names of the most primitive and simple

---

[12] Cf. Frege (1969/1979, 139/128): "[The] task we assign to logic is only that of saying what holds with the utmost generality for all thinking, whatever its subject matter."

[13] In several passages, Frege points out to a reader that his words must not be taken literally as this will lead to misunderstandings (e.g. Frege 1892b/1997, 205/193).

[14] van Heijenoort (1967b), Goldfarb (1979).

[15] At this point, she draws a parallel to Kant's words about logic: "Logic must teach us the correct use of the understanding, i.e. that in which it is in agreement with itself" (Kant 1974, 16). See also Frege (1884, § 105).

[16] Cf. also Alnes (1998, 50) and Goldfarb (1997, 62).

elements of logic is, for instance, problematic since no definitions can be provided of what is simple, according to Frege. The most well-known difficulty that arises when he wants to shed light on the terminology that pertains to logic is connected to the manner in which he draws a distinction between concept and object. Frege clarifies what he means by "concept" by pointing to the logical role of a part of a thought, by saying that a concept is a "nonobject" and that a (first order) concept-name always has an empty place that must be filled with a proper name (that designates an object). He seems to preclude the possibility of making a first order concept the subject of a thought (see Frege 1892b/1997, 200/188). This would also be a consequence of the manner in which he elucidates his conception of 'object'; he tells us that it is the part that does not have a predicative function in the proposition, and that names of objects are saturated (see for instance Frege 1891/1997, 6–8/133–134).

Let us consider an example discussed by Frege that has been in focus in the debate, namely the sentence: "The concept 'horse' is not a concept." We see that there is no way of translating to the purely logical language what the author of this sentence tries to communicate. Frege's approach to difficulties of this kind is to make use of elucidations: sentences and expressions that have a pragmatic and propaedeutic role but no place in the system of science. He tends to speak of them as "Erläuterungen".[17] Frege's use of elucidations that are not possible to express in the logical language is often brought to the fore as paradigmatic in the debate. Let us look at a famous passage from Frege's essay "On Concept and Object":

> I do not at all dispute Kerry's right to use the words 'concept' and 'object' in his own way, if only he would respect my equal right, and admit that with my use of the terms I have got hold of a distinction of the highest importance. I admit that there is a quite peculiar obstacle in the way of an understanding with my reader. By a kind of necessity of language, my expressions, taken literally, sometimes miss my thought: I mention an object, when what I intend is a concept. (Frege 1892b/1997, 192/181)

We notice that Frege says that the expressions he uses when he wants to say something about a concept "miss [his] thought". The latter phrase suggests that the expressions or sentences that he uses when "he intends a concept but mentions an object" do not express thoughts, i.e., that they are nonsensical. When Frege speaks of "a kind of necessity of language", the language in question is, of course, German. What is of significance to our interests here is that the use of the term "concept" that Frege has introduced diverges from how it is used in German; given the grammar of this language, the sentence is well-formed. Looking at a footnote to the main text of "On Concept and Object" will clarify this issue. Frege is here commenting on the "awkwardness of language" that "cannot be avoided" when it comes to saying something about a concept, and he writes:

> A similar thing happens when we say as regards the sentence 'This rose is red': the grammatical predicate 'is red' belongs to the subject 'this rose'. Here the words 'the grammatical predicate is red' are not a grammatical predicate but a subject. By the very act of explicitly calling it a predicate, we deprive it of this property. (Frege 1892b/1997, 197/185, footnote H)

---

[17] Cf. e.g. Frege (1906/1984, 301–302/301).

Frege makes it very clear in this passage, that it is "the very act of calling" the predicative expression a predicate that "deprives it of this property". The example shows that not just any utterance that does not express a thought could serve as an elucidation of the primitive terms of the *Begriffsschrift*. In a certain sense, we could say that Frege takes advantage of the divergence between the use of "concept" in the natural language and the logical distinction between concept and object that he introduces, when a sentence such as "The concept 'horse' is not a concept" is used for elucidatory purposes. Its elucidatory function is connected to our coming to see that the expression is not a meaningful proposition about a concept (i.e., it is not a proposition that expresses a thought). The picture we get from this way of looking at the example, is that Frege enters into nonsensical language for the purpose of elucidating the logical framework within which our rational discourse takes place, and further that bringing others to the insight that what they conceived of as meaningful statements are in fact nonsensical strings of words is a crucial aim that he pursues.[18]

Before moving on to a discussion of Frege's conception of '*Dichtung*', let us look at some further examples of Frege's elucidations. I mentioned earlier that he uses figurative modes of expression for elucidatory purposes, and in the essay "Negation" he writes:

> Figurative expressions [*Bildliche Ausdrücke*], if used cautiously, may after all help towards an elucidation. I compare that which needs completion to a wrapping, e.g. a coat, which cannot stand upright by itself; in order to do that, it must be wrapped round somebody. The man whom it is wrapped round may put on another wrapping, e.g. a cloak. The two wrappings unite to form a single wrapping. There are thus two possible ways of looking at the matter; we may say either that a man who already wore a coat was now dressed up in a second wrapping, a cloak, or, that this clothing consists of two wrappings—coat and cloak. These ways of looking at it have absolutely equal justification. The additional wrapping always combines with the one already there to form a new wrapping. Of course we must never forget in this connection that dressing up and putting things together are processes in time, whereas what corresponds to this in the realm of thoughts is timeless. (Frege 1919/1997, 157/361; modified translation)[19]

Frege describes what he does in this passage as "comparing", and in other passages he tells us that figurative modes of expressions may provide a certain way of looking at something or a certain mode of representation.[20] The metaphors Frege presents are often elaborate (as in the passage quoted above) and he tends to speaks of them as "Gleichnisse", but, as Gabriel has pointed out, he avoids the expression

---

[18] This example connects with what we saw in the quotation from the *Preface* to the *Begriffsschrift*, where Frege suggests that it is a task of philosophy to "break the domination of words over the human spirit by laying bare illusions that through the use of language often almost unavoidably arise". Here, Frege points at a potentially liberating and therapeutic aim that the philosopher might reach that corresponds with the overall picture of similarities between Wittgenstein's *Tractatus* and Frege's work that is presented in the work of Conant and Diamond.

[19] Cf. also Frege (1969/1979, 201/185).

[20] Cf. Frege (1969/1979, 161/149).

"Metapher".²¹ "Gleichnis" does not seem to have a proper counterpart in English, unless we accept "simile" as a translation. The German word is used to refer to a way of using language that belongs in poetic language, but also has an established place in everyday language as well as in scientific investigations. "Simile" is primarily a literary term in English. A comparison is made in a "Gleichnis", but it is not a statement of truth or identity: nothing is asserted by it (at least if we follow Frege's conception of 'assertion' and 'assertoric force').²² In this manner, elucidations in the form of figurative modes of speech seem to offer a mode of expression for Frege's philosophical (conceptual and logical) investigation that makes it possible to avoid the problems that are connected with attempts at providing definitions of primitive elements of logic or at theorizing about this science as if from a standpoint external to it.²³

But let us now recall the picture of Frege's understanding of his purely logical language and its relation to our communication in language that I outlined earlier: The picture tells us that sentences in natural language that cannot be translated to, or expressed in the *Begriffsschrift* don't represent cases of rational discourse. We may consider this picture in connection with the common conception that Frege's elucidations are sentences and expressions that fail to translate into the logical language, and connect this with what we have seen: that Frege often uses poetic forms of language, such as simile and metaphor, for elucidatory purposes. The question that arises is whether it is reasonable to say that the use of figurative forms of language, and in particular metaphor and simile, represent cases of language use that are meaningful according to Frege. That is, whether poetic and literary uses of language can be described as forms of discourse that are possible to understand. I will approach this question by looking at some passages where Frege discusses *Dichtung*.

7.4 Dichtung

Throughout his writings, Frege provides his reader with examples from Homer's epic poems, and it seems that Odysseus was his favorite hero. But Frege also discusses poetry, stage art and fiction more generally and has suggestions to ways of distinguishing more clearly between the use of language within a literary or artistic context, and the use of language in contexts of other kinds. In "On *Sinn* and *Bedeutung*", for example, he discusses our different attitudes when we enquire after

---

[21] It has been suggested by Gabriel that Frege's avoidance of "metaphor" when speaking of his use of figurative language is connected with a wish to put a distance between his work and the work of a poet. It is reasonable to conceive of Frege's choice of words in this case from the perspective of his struggle against psychologism. See Gabriel (1991, 679), and my discussion below.

[22] Cf. Frege's discussion in the essay "My Basic Logical Insights" (Frege 1969/1979, 271–272/251–252), and his discussion of assertoric force in connection with a distinction between a "Hilfssprache" and a "Darlegungssprache" (1969/1979, 281/260).

[23] Cf. Frege (1884, § 88), Frege (1969/1979, 37/33), Gabriel (1991, 71–79), Ricketts (1985), Goldfarb (1979, 353), and Weiner (2001).

the truth of a sentence and when we accept a case of language use as a work of art. He writes: "In hearing an epic poem, for instance, apart from the euphony of the language we are interested only in the sense of the sentences and the images and feelings thereby aroused. The question of truth would cause us to abandon aesthetic delight for an attitude of scientific investigation" (Frege 1892a/1997, 33/157). He goes on to suggest in a footnote that it would be desirable to have a special expression for signs that are meant to have a *Sinn* but no *Bedeutung*. They could, he suggests, be called "images" [*Bilder*]. The words of the actor on stage would be *Bilder*, and the actor himself would in that case be a *Bild* (footnote F).

The discussion in "On *Sinn* and *Bedeutung*" shows that we are, according to Frege (1892a/1997, 33/157), ordinarily interested in reaching the truth and "we generally recognize and expect a *Bedeutung* for the sentence". The language used in fiction or poetry represents, according to this, an exception to what is ordinarily the case, and at times Frege uses forms of expression that may be interpreted as depreciatory of poetry and fiction. He writes, for instance, that:

> Instead of speaking of 'fiction' [*Dichtung*], we could speak of 'sham thoughts' [*Scheingedanken*]. Thus if the sense of an assertoric sentence is not true, it is either false or fictitious [*Dichtung*], and it will generally be the latter if it contains a sham proper name. The art of poetry and literature [*Dichtkunst*], just as, for example, the art of painting, looks to appearances. Assertions in fiction [*Dichtung*] are not to be taken seriously: they are only sham assertions. Even the thoughts are not to be taken seriously as in the sciences: they are only sham thoughts. If Schiller's *Don Carlos* were to be regarded as a piece of history, then to a large extent the drama would be false. But a work of fiction is not meant to be taken seriously in this way: it is all play [*es ist ein Spiel*]. [...] We have a similar thing in the case of historical painting. As a work of art it simply does not claim to give a visual representation of things that actually happened. A picture that was intended to portray some significant moment in history with photographic accuracy would not be a work of art in the higher sense of the word, but would be comparable to an anatomical drawing in a scientific work. (Frege 1969/1979, 141–142/130; modified translation)

The posthumously published essay from which this passage stems was written in 1897, i.e. a few years later than "On *Sinn* and *Bedeutung*". In the latter essay, Frege remarks that "the thought [expressed by the sentence] loses value for us as soon as we recognize that the *Bedeutung* of one of its parts is missing" (Frege 1892a/1997, 33/157). It is a remark that may be read as expressing a depreciative approach to the language of *Dichtung*—poetry, literature and fiction. But in the same essay, Frege provides examples from science of expressions that are used without a *Bedeutung*, such as "the celestial body most distant from the Earth" and "the least rapidly convergent series" (Frege 1892a/1997, 27/153). Frege's remark about the loss of value of a thought for us, when we see that it lacks *Bedeutung*, should hence be seen against the background of a distinction between the goals we pursue in different contexts and activities.

To better see what this perspective involves, we might recall the comparison Frege makes in different passages between the relation between the words "good" and "beautiful" to ethics and aesthetic, and the way the word "true" points the way for logic (for instance in the first lines of the essay "Thought"). It has been suggested by Gabriel (1984) that we should conceive of this comparison as the context for

Frege's introduction of "truth value" as the *Bedeutung* of a sentence, rather than the analogy between a concept, and a function in mathematics. A sentence is valuable, according to this way of looking at it, if it is true or false in a context where truth is strived for. The sentence has significance or importance—*Bedeutung*—when we engage in scientific activities, in so far as it makes sense to pose the question: is it true or false? But when it comes to poetic language, the goal is not truth (in a sense that is relevant to logic) but beauty. The suggested way of looking at Frege's introduction of truth value as the *Bedeutungen* of sentences elucidates remarks such as the following:

> I now ask: does the whole proposition only have a sense, or does it also have a *Bedeutung*? What we talk about is the *Bedeutungen* of words. We say something about the *Bedeutung* of the word 'Sirius' when we say: 'Sirius is bigger than the Sun'. This is why in science it is of value to us to know that the words used have *Bedeutung*. Of course in poetry and legend it makes no difference to us. When we merely want to enjoy the poetry we do not care whether, e.g., the name 'Odysseus' has a *Bedeutung* (or as it is usually put, whether Odysseus was an historical person). The question first acquires an interest for us when we take a scientific attitude—the moment we ask, 'Is the story true?' i.e. when we take an interest in the truth-value. In poetry too there are thoughts, but there are only pseudo-assertions. (Letter to Russell 28.12.1902, in Frege 1997, 256)

In the remark we looked at earlier, Frege told us that in fiction there are only sham thoughts.[24] But in this letter to Russell from 1902, Frege maintains that in poetry too "there are thoughts" and in "On *Sinn* and *Bedeutung*" we find Frege asking:

> Is it possible that a sentence as a whole has only a sense, but no *Bedeutung*? […] The sentence 'Odysseus was set ashore at Ithaca while sound asleep' obviously has a sense. But since it is doubtful whether the name 'Odysseus' occurring therein, has a *Bedeutung*, it is also doubtful whether the whole sentence does. […] The thought remains the same whether 'Odysseus' has a *Bedeutung* or not. (Frege 1892a, 32–33/1997, 157).

If the poet's words about Odysseus express thoughts and the name "Odysseus" has a sense, there is little to support an assumption that *Dichtung* is impossible to understand. In connection with this particular example I would like to return to footnote F in "On *Sinn* and *Bedeutung*" on how it would be desirable to have a special expression for signs that are meant to have *Sinn* but no *Bedeutung*. At this point, I merely want to observe that Frege suggests, in the footnote, that the words of an actor on stage represent or are images of words that are used with both *Sinn* and *Bedeutung*, and that the actor can be seen as an image (or representation) of a person who makes assertions. The name "Odysseus" would accordingly be an image or representation of the name of a (historical) person. But from within the context of the epic poem it is not easy to say what *Sinn* the name has. "Odysseus" seems to function as the name of a person with a specific fate and properties, but we could also say that the *Sinn* of "Odysseus" in the epic poem is *men* in contrast to women or, for instance, *humankind* as opposed to the Gods. It is, I want to suggest, crucial to the context of use here that we are unable to give one single answer to the

---

[24] Cf. Frege (1969/1979, 142/130).

question of what *Sinn* "Odysseus" has in the epic poem.[25] Of course, Frege tells us, there is no single answer to the question of what *Sinn* the name "Aristotle" has either—it may differ from one person to another, but we can tolerate the differences as long as we manage to understand each other (Frege 1892a/1997, 27/153, footnote B). The two examples are however different: the irregular connection between *Sinn* and *Bedeutung* in the latter example is spoken of as a hindrance that should be minimized, but in the case of *Dichtung* the opposite is often the case. Ambiguity is the poet's vehicle and there is no direct analogy between the two cases.

It is clear from these considerations that no question of truth can be raised in the case of representations of this kind if we want to speak of truth in a sense that is relevant to logic, in Frege's view, and further that it is doubtful that the sentence about Odysseus can be translated to the purely logical language (at least from the perspective I have suggested). The question that comes up is what it means to say that a sentence can be expressed in, or translated to the logical language. Is it enough that it has the function-argument structure that Frege points out as fundamental to the logical language? In that case a sentence that has a *Sinn* but no *Bedeutung* can be expressed in the *Begriffsschrift*. Or does the possibility of translation into the *Begriffsschrift* also require that a sentence has both *Sinn* and *Bedeutung*? If we consider the role Frege gives to the horizontal stroke (Frege 1891/1997, 21/142)—a function that takes an object as its argument, and gives *the True* or *the False* as its value—it seems that sentences that have *Sinn* but no *Bedeutung* cannot be expressed in the logical language.

7.5 Philosophy and Understanding

To bring out some connections between Frege's use of elucidatory prose and his conception of 'philosophy' and 'understanding', I will at this point direct attention to two aspects of Frege's discussions. The first concerns the incompleteness of the picture of Frege's elucidations that I have presented so far: several factors indicate that Frege sometimes uses sentences and prose that can be translated to the purely logical language for elucidatory purposes, as e.g. Weiner (2001) and Ricketts (1996) convincingly have argued. Let's look at how Frege's describes the role of elucidation in the essay "Logic in Mathematics":

> Definitions proper must be distinguished from *elucidations* [*Erläuterungen*]. In the first stages of a science [Wenn wir die Wissenschaft beginnen] we cannot avoid the use of ordinary words [die Wörter unserer Sprache]. But these words are, for the most part, not really appropriate for scientific purposes, because they are not precise enough and fluctuate in their use. Science needs technical terms that have precise and fixed *Bedeutungen*, and in order to come to an understanding about these *Bedeutungen* and exclude possible misunderstandings, we provide elucidations. Of course in so doing we have again to use ordinary words, and these may display defects similar to those which the elucidations are

---

[25] The relation between *Sinn* and *Bedeutung* in cases like this (in *Dichtung*) is not as in scientific discourse, but *unscientific*.

intended to remove. So it seems that we shall then have to provide further elucidations. Theoretically one will never really achieve one's goal in this way. In practice, however, we do manage to come to an understanding about the *Bedeutungen* of words. Of course we have to be able to count on a meeting of minds [ein verständnisvollen Entgegenkommen], on other's guessing what we have in mind. But all this precedes the construction of a system and does not belong within a system. In constructing a system it must be assumed that the words have precise *Bedeutungen* and that we know what they are. Hence we can at this point leave elucidations out of account and turn our attention to the construction of a system. (Frege 1969/1979, 224/207)

In this posthumously published essay (written in 1914) Frege observes that we must use ordinary words [die Wörter unserer Sprache] in the beginning of a science. To introduce the technical terminology [Kunstausdrücke] of the science "we provide elucidations", Frege writes. When this is done, ordinary words are used that may have defects that are similar to the ones that the elucidations are to remove. Frege's worry is hence that the words used in elucidations lack a precise meaning. The fact that the procedure he suggests does not rule out failure from a theoretical perspective is pointed out, but Frege argues that it works in practice. He simply bids us to trust in our ability to guess what someone has in mind, and in a "verständnisvolles Entgegenkommen", i.e. an approach of willingness to understand. The German expression (and related ones that Frege uses), is commonly translated by the idiom "a meeting of minds". The translation is, in my view, misleading for several reasons; commentators who relate to this translation, sometimes overlook the metaphorical character of the English idiom and fail to notice that Frege doesn't use the word "mind" [*Geist*] in these passages. If he had done so, and in the plural as in the English idiom, it would have been an indication to the effect that we are here dealing with the states and contents of individual minds. Frege's appeal in passages of this kind would then have been to phenomena that fall entirely within the realm of the subjective. That is, phenomena that do not fall within the realm of what is rational (logical and objective), if "rational" is used in a sense that lies close to "scientific". But Frege's use of the German expression "das entgegenkommende Verständnis" and closely related expressions, instead make it clear that he is in these cases speaking of a form of *understanding*.[26] And the aim of elucidation is here (and elsewhere) presented as the *mutual understanding* of investigators who are using a terminology that pertains to a specific science, such as logic. We can then compare the role of elucidations, to the way in which we in everyday life make certain statements to exclude misunderstandings—as when we say in conversation, for instance: "When I said 'I pray the summer will be full of sunshine', I didn't mean that in a religious or metaphorical sense. I simply expressed my wish for nice weather during summer."[27] This comparison agrees with Frege's words in a letter to Hilbert from 1899 where he speaks of "elucidatory propositions" as similar to definitions (that

---

[26] One might compare Frege's appeal in these cases to Kant's notion of *sensus communis*, and his distinction between *sensus communis aestheticus* and *sensus communis logicus* in the Third Critique, § 40. Cf. also Cavell's implicit reference to this Kantian notion in his famous essay "Aesthetic Problems of Modern Philosophy" (Cavell 2002, 73–96).

[27] Cf. Ishiguro (1969, 32–33) on Wittgenstein's conception of 'elucidation' in the *Tractatus*.

are different from axioms and theorems) in that "they too are concerned with laying down the meaning of a sign (or word)." He goes on to say that they are used when one is concerned with "warding off the unwanted meanings among those that occur in linguistic usage and [...] pointing to the wanted one" (Frege 1980, 36).

The second aspect that I want to draw attention to is Frege's reliance on our competence as speakers of a first language when he offers elucidations in the form of figurative expressions. He directs our attention to this reliance in various places, for instance when he writes that he must rely on a reader's "feeling for the German language" and that he must trust a reader to take his hints in the right manner and get that his words are not to be taken in a literal way—otherwise his hints will not reach their purpose.[28] The ability to take a metaphor or simile in the right manner certainly presupposes that we are competent speakers of the relevant language. To clarify this point we may again look at the famous passage in "On Concept and Object" where Frege discusses the clarification of the logical distinction between 'object' and 'concept'. He writes:

> I admit that there is a quite peculiar obstacle in the way of an understanding with my reader. By a [certain] kind of necessity of language, my expressions, taken literally [*ganz wörtlich genommen*], sometimes miss my thought; I mention an object, when what I intend is a concept. I fully realize that in such cases I was relying upon a reader who would be ready to meet my half-way [*auf ein wohlwollendes Entgegenkommen des Lesers*]—who does not begrudge me a pinch of salt. (Frege 1892a/1997, 204/192)

We notice how Frege points out that it is when his expressions are "taken literally" that they "sometimes" miss his thought. He could hence be read as saying that when he wants to clarify how he uses the term "concept", he *sometimes* uses expressions that do not express thoughts (in a technical/logical sense of "thought"). But we could also understand him in the following way: *if* the expressions he uses are taken in a literal sense, they will "miss his thought" (in a non-technical sense of "thought").[29] Frege hereby indicates that taking his words literally involves a mistake. If they are not so taken—if our approach is a willingness to understand—his words will perhaps not fail to indicate what he is after, we might grasp the relevant distinction. The grasping or understanding is not a matter of individual minds sharing a common content or state— instead, we grasp something objective.

In this sense we could say that Frege's scientific project can be successful only given our understanding of the language of life [*die Sprache des Lebens*] (where communication is characterized by guesswork, imprecise terminology, insufficient expressions of thoughts, fluctuation in the use of words and where an ability to understand body language and gestures is vital, according to Frege). Our ability to understand poetic language in the right way (as a form of discourse where words and sentences have *Sinn* but often lack *Bedeutung*) can then also be described as a presupposition for the success of Frege's project. We must be able to see when we are dealing with sentences or words that are not put forth in a literal mode, and understand that there can be no paraphrase to literal statements of the metaphors

---

[28] Cf. Frege (1892b/1997, 195, 205/184, 193), (1893, 4), (1969/1979, 254/235).
[29] Cf. here Frege (1918/1997, 60–61/327–328).

that have been used. Let us return to the passage from "Thought" that we first looked at:

> On the other hand, the constituents of language to which I here want to call attention make the translation of poetry [*Dichtungen*] very difficult, indeed make perfect translation almost always impossible, for it is just in what largely makes the poetic value [*dichterische Wert*] that languages most differ. (Frege 1918/1997, 63/330)

I said that Frege's project can only be successful given our understanding of the language of life. If this means that Frege must rely on his readers to be language users, or presuppose that people communicate in language, it's an empty requirement. But the point I wanted to make can be seen in Frege's directing our attention to differences between languages in the quoted passage from the essay "Thought". Frege's remark on poetic value and the difficulty of translation of *Dichtungen* connects with the features of Frege's elucidations that I have wanted to bring to the forth, and underscores the importance of his reader's ability to identify and interpret sentences and prose of this kind for the success of his project. If my suggestions are reasonable, the picture of how we can use Frege's *Begriffsschrift* to test whether an utterance is an instance of rational discourse falls apart—at least if we use "rational" in a wider and more general sense than the one that singles out assertions by which judgments are manifested as cases of rational discourse. It further seems reasonable to say that an attempt to tie Frege to a purely scientific conception of philosophy will fail as we consider his outline of the distinction between the humanities and the exact sciences in the essay "Thought" (Frege 1918/1997, 63/330). It has become clear that in his philosophical writings, Frege often moves in a field that lies closer to the first than to the second kind of science, in the sense that he aims at elucidating what cannot be conceptually grasped, by way hints—*auf dem Wege der Ahnung*[30].

References

Alnes, J. H. (1998). *Frege on logic and logicism*. Oslo: Acta Humaniora.
Cavell, S. (2002). *Must we mean what we say?: A book of essays*. Cambridge: Cambridge University Press.
Conant, J. (2000). Elucidation and nonsense in Frege and early Wittgenstein. In A. Crary & R. Read (Eds.), *The new Wittgenstein* (pp. 174–217). London: Routledge.
Diamond, C. (1991). *The realistic spirit: Wittgenstein, philosophy and the mind*. Cambridge, MA: MIT Press.
Floyd, J. (1998). Frege, semantics, and the double definition stroke. In A. Biletzki & A. Matar (Eds.), *The story of analytic philosophy: Plot and heroes* (pp. 141–166). London: Routledge.

---

[30] I would like to thank Jan Harald Alnes for helpful suggestions and very valuable comments to drafts for this paper. I'm also grateful for comments and suggestions from participants at the workshop "Frege zwischen Dichtung und Wissenschaft" in Bergen, and from participants of the research seminars at Uppsala University, the University of Oslo and Tampere University. In particular, I would like to thank Joan Weiner, Juliet Floyd, Leila Haaparanta, Mirja Hartimo and Sharon Rider. Work on this paper was financed by the Research Council of Norway.

Frege, G. (1879). *Begriffsschrift, eine der arithmetischen nachgebildete Formelsprache des reinen Denkens*. Halle: Louis Nebert.
Frege, G. (1882). Ueber die wissenschaftliche Berechtigung einer Begriffsschrift. *Zeitschrift für Philosophie und philosophische Kritik, 81*, 48–56.
Frege, G. (1884). *Grundlagen der Arithmetik: Eine-logisch mathematische Untersuchung über den Begriff der Zahl*. Breslau: Verlag von W. Koebner.
Frege, G. (1891). *Function und Begriff: Vortrag, gehalten in der Sitzung vom 9. Januar 1891 der Jenaischen Gesellschaft für Medizin und Naturwissenschaft*. Jena: Hermann Pohle. English edition: Frege, G. (1997). Function and concept. In G. Frege, *The Frege reader* (pp. 130–148). M. Beaney (Ed.). Oxford: Basil Blackwell.
Frege, G. (1892a). Über Sinn und Bedeutung. *Zeitschrift für Philosophie und philosophische Kritik, 100*, 25–50. English edition: Frege, G. (1997). On Sinn and Bedeutung. In G. Frege, *The Frege reader* (pp. 151–172). M. Beaney (Ed.). Oxford: Basil Blackwell.
Frege, G. (1892b). Über Begriff und Gegenstand. *Vierteljahrsschrift für wissenschaftliche Philosophie, 16*, 192–205. English edition: Frege, G. (1997). On concept and object. In G. Frege, *The Frege reader* (pp. 181–193). M. Beaney (Ed.). Oxford: Basil Blackwell.
Frege, G. (1893/2013). *Grundgesetze der Arithmetik: Begriffsschriftlich abgeleitet, I. Band*. Jena: Verlag von Hermann Pohle.
Frege, G. (1906). Über die Grundlagen der Geometrie [second series]. *Jahresbericht der Deutschen Mathematiker-Vereinigung, 15*, 293–309 (Part I), 377–403 (Part II), 423–430 (Part III). English edition: Frege, G. (1984). On the foundations of geometry: Second series. In G. Frege, *Collected papers* (pp. 293–340). B. McGuinness (Ed.). Oxford: Basil Blackwell.
Frege, G. (1918). *Der Gedanke: Eine logische Untersuchung. Beiträge zur Philosophie des deutschen Idealismus*, I, 58–77. English edition: Frege, G. (1997). Thought. In G. Frege, *The Frege reader* (pp. 325–345). M. Beaney (Ed.). Oxford: Basil Blackwell.
Frege, G. (1919). Die Verneinung: Eine Logische Untersuchung. *Beiträge zur Philosophie des deutschen Idealismus, I, 2*, 143–157.
Frege, G. (1969). *Nachgelassene Schriften*. H. Hermes, F. Kambartel, &, F. Kaulbach (Eds.). Hamburg: Felix Meiner Verlag. English edition: Frege, G. (1979). *Posthumous writings*. H. Hermes, F. Kambartel, & F. Kaulback (Eds.), P. Long, & R. White (Trans.). Chicago: University of Chicago Press.
Frege, G. (1980). *Philosophical and mathematical correspondence*. G. Gabriel, H. Hermes, F. Kambartel, C. Thiel, & A. Veraart (Eds.), H. Kaal (Trans.). Chicago: University of Chicago Press.
Frege, G. (1997). *The Frege reader*. M. Beaney (Ed.). Oxford: Blackwell.
Friedman, M. (1992). *Kant and the exact sciences*. Cambridge, MA: Harvard University Press.
Gabriel, G. (1984). Fregean connection: Bedeutung, value and truth-value. *The Philosophical Quarterly, 34*(136), 374–375.
Gabriel, G. (1991). *Zwischen Logik und Literatur: Erkenntnisformen von Dichtung, Philosophie und Wissenschaft*. Stuttgart: Metzler.
Gabriel, G., & Polimenov, T. (2012). Analytical philosophy and its forgetfulness of the Continent: Gottfried Gabriel in conversation with Todor Polimenov. *Nordic Wittgenstein Review, 1*(2012), 155–177.
Glock, H.-J. (2008). *What is analytic philosophy?* Cambridge: Cambridge University Press.
Goldfarb, W. (1979). Logic in the twenties: The nature of the quantifier. *The Journal of Symbolic Logic, 47*, 351–368.
Goldfarb, W. (1997). Metaphysics and nonsense: On Cora Diamond's *The Realistic Spirit*. *Journal of Philosophical Research, 22*, 57–73.
Ishiguro, H. (1969). Use and reference of names. In P. Winch (Ed.), *Studies in the philosophy of Wittgenstein* (pp. 20–50). London: Routledge and Kegan Paul.
Kant, I. (1974). *Logic*. L. Hartman & W. Schwartz (Trans.). Indianapolis: Bobbs-Merrill.
Quine, W. V. O. (1981). *Theories and things*. Cambridge, MA: Harvard University Press.
Ricketts, T. (1985). Frege, the *Tractatus*, and the logocentric predicament. *Noûs, 9*, 3–15.

Ricketts, T. (1996). Logic and truth in Frege: I. *Proceedings of the Aristotelian Society Supplementary Volume, 70*, 121–140.
van Heijenoort, J. (Ed.). (1967a). *From Frege to Gödel*. Cambridge, MA: Harvard University Press.
van Heijenoort, J. (1967b). Logic as language and logic as calculus. *Synthese, 17*, 324–330.
von Wright, G. H. (1994). Analytische Philosophie: Eine historisch-kritische Betrachtung. In G. Meggle & U. Wessels (Eds.), *Analyomen 1, proceedings of the 1st conference 'Perspectives in Analytical Philosophy'* (pp. 3–30). Berlin: Walter de Gruyter.
Weiner, J. (1990). *Frege in perspective*. Ithaca: Cornell University Press.
Weiner, J. (2001). Theory and elucidation: The end of the age of innocence. In J. Floyd & S. Shieh (Eds.), *Future pasts: The analytic tradition in twentieth century philosophy* (pp. 43–66). New York: Oxford University Press.
Weiner, J. (forthcoming). Frege's unmetaphysical story about natural language and truth.
Wittgenstein, L. (2001). *Philosophical investigations/Philosophische Untersuchungen* (3rd ed.). G. E. M. Anscombe (Trans.). Oxford: Basil Blackwell.

Chapter 8
Semantic and Pragmatic Aspects of Frege's Approach to Fictional Discourse

Todor Polimenov

Abstract In many places in his works Frege comes to speak of fiction. Sometimes he appeals to it to get the background against which to draw the semantic boundaries of his logical investigations. Sometimes he gives examples from fiction to clarify some specific relations between his semantic concepts. It is worth analyzing Frege's remarks on fiction in order to see if they contain insights that let us elaborate a Fregean definition of fictional discourse. It is shown that they not just negatively say what fictional discourse is not, but also do indicate what it is. Furthermore, it is important to distinguish between semantic and pragmatic features of Frege's view of fiction. The pragmatic ones, it is argued, anticipate some basic insights of a speech-act theoretical approach to fictional discourse. In addition the paper explores what Frege would tell us about the ontological status of fictional objects if the truth conditions of statements about them are taken into consideration in a Fregean manner.

Keywords Fictional discourse • Fictional objects • Frege • Illocutionary force • Reference • Sense

8.1 Introduction

Frege's research interests lay within the areas of logic and philosophy of mathematics. In search of a new logical theory for the purposes of his mathematic-logical program Frege found himself forced to deal with language firstly. The analysis of language, or more precisely, of the way we talk about the world is essentially the method by which he rejects a number of traditional concepts and introduces completely new distinctions and directions in logic. As a result, what turned Frege into one of the founding fathers of the philosophy of language in the analytic tradition were the fundamental insights into the working of language he had gained during

T. Polimenov (✉)
Sofia University, Sofia, Bulgaria
e-mail: todor.polimenov@phls.uni-sofia.bg

his logical-linguistic analyses and, more generally, his original approach to the questions of logic from the point of view of the question of the 'sense' of our statements. Thus, although for him investigations into language had never been an independent objective of its own, today he is considered one of the key figures in the philosophy of language.

Even less did Frege ever intend to develop a theory of fictional discourse. He, as it seems, discusses fiction and the poetic use of language only peripherally in his works and the respective considerations rarely amount to more than fragmentary remarks. On closer examination, however, one notices a few significant details. Firstly, the reference to fictional discourse gives Frege an important background against which to draw the semantic boundaries of his own logical investigations. Secondly, Frege repeatedly gives examples from fiction at crucial junctures in his writings, where specific relations between his semantic categories are being explained. Thirdly, without trying to define fictional discourse—assuming that we have already mastered it—Frege points to certain features that distinguish it from other ways of speaking and that can be used to characterize it. That is why a study of Frege's approach to fiction appears to promise to shed valuable light both on Frege's philosophy of language and on the very nature of fictional discourse.

In order to get a fuller picture of Frege's views on fictional discourse and to make those views more usable for contemporary debates, we need to—apart from exploring the contexts in which Frege explicitly deals with fiction—also consider the question how, with the help of the tools we inherited from Frege (namely the distinctions he had drawn and the concepts he had got on their basis), we could analyze certain kinds of fictional statements that Frege did not discuss. It is those kinds of statements that are often examined in the secondary literature to test a given theory of fiction of a 'Fregean' type. For that reason, they seem to be a good means for assessing the explanatory potential of such a theory.

Certainly, we do not have a full-fledged theory of fictional discourse in Frege. On the other hand, Frege worked out in detail enough semantic distinctions through which a theory of the kind can be developed. Let us mention the most important ones: the distinction between sense (*Sinn*) and reference (*Bedeutung*), the distinction between customary and indirect reference (more generally, in Carnap's idiom: between extensional and intensional contexts), the distinction between thought and assertoric force (more generally, in Searle's idiom: between propositional content and illocutionary force), the distinction between assertion (*Behauptung*) and pseudo-assertion (*Scheinbehauptung*). This is the reason why in the present paper I speak only of Frege's *approach* to and not of a *theory* of fictional discourse.

The first commentator to notice that Frege's semantics contains elements that can be employed in the theory of fiction was, it seems, Aschenbrenner (1968). His research, however, remains limited to general suggestions of how to apply the distinctions from "Über Sinn und Bedeutung" (*SB*) to explain the 'ontology' of literary works of fiction. He neither took into consideration other passages from Frege's works relevant to the topic as, for instance, those in "Der Gedanke", nor—which is more important—could he have used texts from the Frege *Nachlass* that reveal a number of additional aspects of Frege's views on fiction.[1] A decisive influence on the further discussions was Searle (1975/1979). Although Searle's exposition is systematic and Frege's name is never mentioned, Searle offers an account of fictional discourse, which at least gives the impression of being a development of Fregean ideas placed within a larger speech-act theoretical framework. Whether Searle, of whom we know that he had at an earlier stage studied Frege's semantics, was directly influenced by it in that particular case is not important here. What matters is that his speech-act theoretical perspective on fictional discourse gives us the opportunity to see Frege's own remarks on fiction in a new light. Thus, we can discover in them some features that, today, we can call pragmatic, and claim that Frege at least had anticipated some of Searle's central tenets.

Besides the question of the nature of fictional discourse (the topic of speaking *in* works of fiction) Frege's semantic distinctions have their contemporary relevance mostly to discussions of the ontological status of fictional objects (the topic of speaking *about* fictional works and their characters). A considerable challenge to Frege scholars is presented by the neo-Meinongian approach of Parsons (cf. 1975 and particularly the systematic exposition in 1980); especially in that Parsons (1982) targets directly "Fregean theories of fictional objects" in order to show that they create rather than solve problems with the analysis of sentences which seem to support the assumption that *there are*—in some way—objects that *do not exist*. Examples of such sentences are:

(1) Sherlock Homes is a fictional detective who is more famous than any real detective, living or dead.
(2) A certain fictional detective is more famous than any real detective.[2]

Of course, responses of Fregeans have not taken long to appear (cf. Gabriel 1987/1993; Künne 1995), and debates concerning Frege's conception of fiction and the application of his distinctions to its explanation are still underway (Textor 2011 is representative of the current state of the discussion).

8.2 Fiction and *Dichtung*

The central notion in Frege's semantics is the notion of truth (Frege 1897/1979, 128 ff., 1918/1984, 351 ff., 1919b/1979, 253). According to Frege, it is indefinable. Since truth is a primary phenomenon, the notion of it cannot be decomposed into logically simpler elements and thus defined. That is why we have to proceed from it and use it as a guide for further logico-semantic distinctions. How do we know,

---

[1] The first volume of Frege's *Nachgelassene Schriften* was published a year later, and his *Wissenschaftlicher Breifwechsel* in 1976 (cf. the English translations in Frege 1979 and 1980).

[2] (2) is especially intriguing, because according to modern logical analysis, (2) seemingly expresses an existential statement: "$\exists x\, (x$ is a fictional detective $\land\, \forall y\, (y$ is a real detective $\rightarrow x$ is more famous than $y))$".

however, that a phenomenon has to do with truth at all? In that respect we can only note: truth is expressed in a speech act of asserting, in uttering a sentence with assertoric force (Frege 1915/1979, 252, 1918/1984, 356). Whoever masters the practice of making assertions already knows what it means to make a truth-claim. It is therefore not surprising that the topic of fictional discourse surfaces in Frege's works precisely in connection with the notion of truth.

When speaking of fiction, Frege uses the German word *Dichtung* (or the derivatives: "dichterische Sprache", "dichterischer Gebrauch", "Dichtkunst", "Dichter"). Besides 'fiction', *Dichtung* can also mean 'poetry', 'poem' (and, respectively, its derivatives can mean: 'the language of poetry', 'poetic use', 'poet', etc.). Probably, Frege uses *Dichtung* in order to sharpen the contrast between fictional discourse, which is, so to say, truth-free, and the language of sciences through which their aim is pursued: the discovery of truths.

Frege speaks of *Dichtung* in two different thematic contexts. The first one concerns truth-valueless sentences—sentences that, although expressing thoughts, belong to the realm of fiction. The other context concerns words or means of expression through which the so-called 'coloring' ('Färbung') of thought is achieved (Frege 1892a, 161, 1897/1979, 139, 1918/1984, 356 ff.). These words are those parts of a sentence, whose meanings do not affect the truth-value-capable core of the sentence's content (the thought). When the role of certain linguistic units is to serve for aesthetic delight by giving speech euphony (rhythm, melody) or by evoking certain feelings and phantasies (*Vorstellungen*) in the mind of the listener or the reader, the translation of *Dichtung* as "fiction" seems inappropriate. Therefore, in these contexts English translations opt for "poetry" etc. I will not deal with the poetic use of language here, but only with fictional discourse. Neither will I discuss figurative speech (the use of metaphors, similes and other rhetorical figures) as it can be encountered both in fictional and in non-fictional discourse (metaphors, e.g., are also used in the language of science, cf. Searle 1975/1979, 60; especially Gabriel 1991).

Let us, as a point of reference, sketch a preliminary definition of fictional discourse, based on views to be found in Frege's works. When we utter an assertoric sentence under ordinary circumstances, we make a claim: things in the world are the way we say they are. This can be called a truth-claim: it is *true* that things in the world are this or that way. Now, whoever utters an assertoric sentence in fictional speech behaves *as if* he makes a truth-claim (Frege 1897/1979, 130, 1918/1984, 356, 1919b/1979, 251). In that case we know: he does not commit himself to the truthfulness of what he says. While in ordinary use the semantic role of assertoric sentences consists in (i) expressing thoughts *and* (ii) claiming that these thoughts are true, uttering assertoric sentences in fiction modifies the semantic rules in such a way that sentences still express thoughts but these thoughts are not put forward as true, it is only pretended to be so.

This affects the semantic rules of the parts of the sentence as well. When someone uses a (simple) assertoric sentence to make an assertion, we *presuppose* that he refers to certain objects in the world through the proper names or definite descriptions that occur in the sentence (otherwise the sentence could be neither true, nor false). The circumstance that the speaker in fictional discourse only behaves as if he asserts something also gives him the opportunity to only behave *as if* he refers to something through the proper names or definite descriptions that he uses (Frege 1892b/1979, 122). Therefore, the self-evident presupposition (*selbstverständliche Voraussetzung*) of ordinary assertoric speech that the objects that are spoken of exist—i.e., that the used proper names and definite descriptions are not empty—is suspended in fictional discourse (cf. Frege 1892a/1984, 168).

We can witness a similar thing with predicates as well. If their semantic role consists of classifying objects, which happens only when they yield criteria by which it *can* be determined for any arbitrary object whether the predicate is true of it or not, then we could say that it would be sufficient for the fictional use of a predicate that it has a sense (that it contributes to the expression of thought) though it may not contain clear principles of classification. Let us take as an example the word "moly" from the *Odyssey* (Frege 1892b/1979, 122). According to Homer's story, moly is a magical herb that has a milky-white flower and a black root. This description familiarizes us with some properties that an herb should have to be a moly and we can say that the word "moly" includes them in its sense as characteristic features (*Merkmale*). These features, however, would not suffice to establish for any object whether it would be true of it that it is a moly or not. Hence, the word "moly", as we know it from the *Odyssey,* is not suitable for proper classification of objects. That does not prevent it from having a sense that makes it possible for us to understand those sections of the *Odyssey* that contain "moly", for instance, the episode in which Hermes gives Odysseus a moly to protect him from Circe's magic. On analogy with complete assertoric sentences and singular terms such as proper names and definite descriptions we can say: anyone who uses a predicate in fictional sentences only behaves *as if* he classifies objects with it.

Our preliminary definition is as follows: fictional discourse is an as-if-speech. Frege confines himself only to assertoric speech because it stands in immediate relation to the question of truth but, certainly, in a work of fiction not only assertions are made, but also questions are asked, commands given, suppositions expressed, etc. With respect to the speech acts of asking questions, giving commands, etc. we can say something similar to what we observed with the speech act of asserting: in fiction, the questions are as-if-questions, commands are as-if-commands and so on.

8.3 Thoughts Being Neither True, Nor False

Once we have already at our disposal a preliminary guide to what fictional discourse is, let us consider in more detail how the question of fiction (*Dichtung*) makes its appearance in Frege's works. Frege starts to refer to fictional discourse only after he has semantically specified his logic through, firstly, a strict distinction between the sign (*Zeichen*) and the thing signified (*Bezeichnetes*), and next—which is more important for the present topic—through distinguishing two components in the 'semantics' of every logically relevant sign: sense and reference. Before that, just once in the *Grundlagen der Arithmetik*, when discussing the domain of geometrical truths, Frege says that in so far as they are intuitable (*anschaulich*), even the wildest phantasies in myths and fiction—phantasies that make "animals talk, ... stones turn into people, and people into trees"—are subject to the laws of geometry (Frege 1884/1950, § 14).

Since the topic of the sense and reference of a sign (singular term, predicate, sentence, logical particle) provides the framework within which Frege discusses fiction, let us describe the semantic categories 'sense' and 'reference' in the following general way: the sense of a sign is what we grasp when we *understand* the sign; reference is what we *mean* by that sign. The sign *expresses* its sense and *designates* its reference, by which the sense mediates the relation between the sign and its reference (the sense presents the reference in a certain way and from a certain perspective). Or in more detail: in expressing a sense, the sign designates the referent determined by that sense. The following scheme can visualize this:

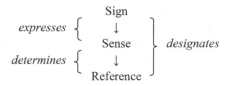

The one and the same reference can be presented by different senses (namely, from different perspectives), and the one and the same sense can be expressed by different signs.

In the context of the distinction between sense and reference the topic of fiction is mentioned for the first time in a letter to Husserl from 24/05/1891 (Frege 1891/1980). Here Frege discusses the reference type of predicates (*Begriffswörter*) and distinguishes it from the type of the references of singular terms (*Eigennamen*), pointing out that whereas for the fictional use of language (*den dichterischen Gebrauch*) it suffices that signs have senses, the scientific use requires their references as well.[3]

The first place where the topic of fiction is considered in more detail and where it plays a key role in the course of the argumentation is *SB*. Here Frege (1892a/1984) argues for two theses: (i) that proper names (and more generally: singular terms)

---

[3] Following the traditional distinction between the content and the extension of a concept, Husserl is inclined to think that what a general term (*Gemeinname*) refers to is a set of objects in a similar way that a singular term refers to one definite object. Frege (1891/1980, 63) objects that a general term—unlike an empty singular term—can have a scientific use even when no object falls under the concept it denotes (when the respective set is empty). Let us take "a satellite of Venus" as an example: the sentence "Venus has no satellites" ("There is no thing that is a satellite of Venus") does not only have a sense but also states a truth. We do not refer to sets by predicates ("... is a satellite of Venus") but by abstract singular terms: "the set of Venus's satellites". The reference of a predicate is a function that maps objects (references of singular terms) to truth-values (references of sentences). Frege calls that function "concept". That is why the concept is not the sense but the reference of a predicate. The sense is the perspective in which the concept is given. For example, the predicates "... is red", "... has the colour with the longest wave length" and "... has the colour of blood" have different senses yet identical reference.

have not only references but also senses; and (ii) that sentences (more precisely: assertoric sentences) have not only senses but references as well. It is in the transition from (i) to (ii) where the topic of fiction becomes part of the argumentation that motivates the introduction of truth-values as references of sentences (cf. also Frege 1902/1980, 152).

The argumentation is as follows. Singular terms have sense and reference (which has already been demonstrated), but do complete sentences have senses and references? They express thoughts (which is obvious) but is the thought their sense or their reference? Since the thought stands in functional dependency to the sense of the singular term but not to its reference—as can be seen from the fact that the substitution of co-referential singular terms in a sentence changes the thought that this sentence expresses, whereas the substitution of synonymous singular terms does not—the thought should be regarded rather as the sense of the sentence. Does the sentence then have a reference too? We know that there are singular terms that have senses but not references. Such are, e.g., proper names in *fiction*. We understand the sentence:

(3) Odysseus was set ashore at Ithaca while sound asleep.

We grasp its sense. It expresses a thought. The name "Odysseus" also has a sense, because if it did not, (3) would contain a part that has no sense and then the whole sentence would also be without sense (cf. also Frege 1914b/1980, 79). Most probably, however, "Odysseus" does not have a reference, which is not a problem in itself, in so far as we are in the domain of fiction. No one expects that fictional names designate real objects. On the other hand, it seems natural to suppose that the reference of a sentence—if sentences have references—will be something that stands in a certain relation to the references of its parts. So, are there cases in which we require of a name to have not only sense but a reference as well? When do we ask, e.g., whether Odysseus did exist? Questions like the last one do not arise in fiction but when we start to take an interest in, e.g., whether the story sentence (3) is a part of is true. Now, the truth-value of a predicative sentence directly depends on which object we apply the predicate to. If such an object does not exist, the sentence cannot have a truth-value. Just as a sentence does not have a sense if one of its parts lacks sense, a sentence does not have a truth-value if one of its parts doesn't have a reference.

By posing the question of truth, however, says Frege, we leave the realm of fiction and no longer consider (3) as part of a fictional story. We start to take interest in facts—in whether (3) depicts truthfully a certain state of affairs.

In this context Frege calls attention to a peculiarity that Searle (1975/1979, 60) uses later on to draw a distinction between ordinary speech and fictional discourse. Whoever commits to the truth of (3), utters (3) *seriously*. Contrariwise, recognizing (3) as part of fictional speech, we know that it should not be taken seriously: here uttering (3) is just a game. In a fictional context by uttering (3) we do not say anything true or false but rather we express a thought that only *seemingly* (*scheinbar*), within the limits of this game, is posited as true (cf. also Frege 1897/1979; 1918/1984).

In a paper draft that further develops the topic of sense and reference (in *SB* they are discussed only for the case of singular terms and sentences) by explaining the distinction between sense and reference with respect to predicates (so that it is not confused with the traditional distinction between the content [*Inhalt*] and the extension [*Umfang*] of a concept, cf. note 3), Frege (1892b/1979, 118, 122) brings up the subject of fiction again. This is the only place where in addition to the usual examples of reference-less singular terms (like the proper name "Nausicaa") Frege also mentions the use of predicates in fiction giving the example with the above-cited word "moly". The point is the same as the one made in *SB*: when we are interested in truth, we ask about the references of sub-sentential expressions (or presuppose that they have references) and we are not contented with the senses of words and sentences as in fiction (cf. also Frege 1904/1980, 165).

8.4 Assertions Without Assertive Force

The paper draft "Logic" from 1897 reveals some new aspects of Frege's view on fictional discourse. The text contains the most extensive discussion of fiction in Frege's works. It is also interesting because it displays, for the first time, the main line along which Frege unfolds his semantic distinctions (in exactly the opposite direction to the one adopted in *SB*): he starts with the notion of truth, introduces the sentence and the thought (the sense of a sentence) as the entities about which the question of truth arises and only then defines the sense of sub-sentential components as their contribution to the expression of thought to finally draw the distinction between sense and reference—not without making a comparison between "serious" and truth-directed speech with fictional speech, which has sense but contains reference-less names and therefrom truth-valueless sentences.

The topic of fiction is introduced in the following way (Frege 1897/1979, 129 ff.). A thought *proper* (*eigentlicher Gedanke*) is always true or false. There is, however, a third case, namely, that of pseudo-thoughts (*Scheingedanken*) which are neither true, nor false. Such thoughts are thoughts in fiction (cf. also Frege 1906b/1979, 198; 1914b/1980, 79 ff.; 1919a/1984, 373 and 389; 1923/1984, 394). Since Frege explicitly calls these pseudo-thoughts "fiction", we can use his criterion for a pseudo-thought in order to define fiction:

(F) The sense of a sentence is fiction if that sentence contains a pseudo-proper name (*Scheineigenname*).

A pseudo-proper name is a name which has a sense but no reference. Reference-less proper names (more generally: reference-less singular terms) seem to deprive the sentences that contain them of a truth-value because the truth-value is nothing else but the sentences' reference. As we said above: if a sentence contains a *part* that has no reference, it seems natural that the sentence itself cannot have one.[4] In this manner neither the sentences:

---

[4] In a footnote Frege (1897/1979, 130) explicitly excludes in this regard the cases in which a pseudo-proper name occurs in indirect speech (in intensional contexts). As we shall see later, that

(4)   Scylla has six heads

or

(5)   William Tell shot an apple off his son's head

nor their negations could be true, because the names "Scylla" and "William Tell", while having sense, do not have reference. Consequently, (4) and (5) are fictional (cf. also Frege 1914a/1979, 232).

Let me call (F) a semantic criterion of fiction. Drawing on the semantic property of a singular term of (having only a sense but) lacking a reference, (F) formulates a sufficient (though not a necessary) condition for something to be fiction (for which sentences are to be characterized as fictional). This criterion has puzzled many Frege interpreters because according to (F) it turns out that, e.g.,

(6)   Vulcan causes disturbances in the orbit of Mercury

and

(7)   The author of the *Principia Mathematica* found an antinomy in the logical system of Frege's *Grundgesetze der Arithmetik*

are fictional sentences. If (F) were adequate, then a substantial part of the history of science would turn out to be fiction, which means: it would consist of sentences that are neither true, nor false.

In the text that immediately succeeds the above discussed one, however, Frege develops his reasoning on fiction as follows:

> The writer, in common with, for example, the painter, has his eye on appearances [Schein]. Assertions in fiction are not to be taken seriously [ernst]: they are only pseudo-assertions [Scheinbehauptungen]. Even the thoughts are not to be taken seriously as in the sciences: they are only pseudo-thoughts. If Schiller's *Don Carlos* were to be regarded as a piece of history, then to a large extent the drama would be false. But a work of fiction is not meant to be taken seriously in this way at all: it's all play [ein Spiel]. Even the proper names in the drama, though they correspond to names of historical personages, are pseudo-proper names; they are not meant to be taken seriously in the work. We have a similar thing in the case of an historical painting. As a work of art it simply does not claim to give a visual representation of things that actually happened. A picture that was intended to portray some significant moment of history with photographic accuracy would be not a work of art in the higher sense of the word, but would be comparable rather to an anatomical drawing in a scientific work. (Frege 1897/1979, 130. In part my translation.)

Frege gives us here a second criterion of fiction I would like to call pragmatic since it concerns the *use* of proper names and sentences. Whether a given instance of speech is fictional or not cannot be decided on the basis of the semantic (or syntactic) properties of the sentences it consists of and of the ones of their parts. It is *shown* in the context of its use (on the condition, of course, that the intentions of the speaker are recognizable). The example of Schiller's *Don Carlos* illustrates the point. We can use the proper name "Don Carlos" to refer to the son of King Philip II of Spain in order to describe certain historical facts, or we can, as Schiller does, use that name to

---

will carry significance in the account of the truth of sentences *about* fictional works and characters.

only pretend to refer to the son of the Spanish king and thus to make up a story that does not bear any or almost any relation to reality. Then that story would consist of sentences by which the storyteller only pretends to make assertions. So, what he makes thereby can be called "pseudo-assertions". In such a use of names and sentences we know: the question of truth does not arise here. This is only a game.

Just as a work of fiction, e.g., Schiller's *Don Carlos,* can be confused with a historical treatise, so, conversely, a well-written historical treatise can be easily staged as a theatrical play (cf. Searle 1975/1979, 59; Gabriel 1983/1991, 7). If that is done, however, we already know: its sentences are not to be taken seriously. Which means: in this context it is irrelevant whether the truth conditions of sentences and, respectively, the reference conditions of proper names and definite descriptions have been fulfilled.

There are literary works which, in addition to fictional proper names such as "Jean Valjean" and "Sherlock Holmes", also contain non-fictional proper names such as "Paris", "Napoleon", "London", "Baker Street" referring, obviously, to real objects. There are novels in which fictional names are used instead of non-fictional ones in order again to refer, as it seems, to real objects (e.g., "Ebenezer Scrooge" in Dickens's *Christmas Carol* is used to refer to the politician John Elwes, or "the Bulgarians" in Voltaire's *Candide* to the Prussians). Can we say that such works of fiction, containing as they do truth-valueless sentences can nevertheless convey some knowledge, that we can, after all, learn something about history and geography from them? Perhaps only a part of a literary work is fictional while another part tells certain truths? This supposition appears to be confirmed by the fact that the authors of literary works often insert universal statements about man and the world among singular sentences about people, places, and events (e.g., "There is nothing either good or bad, but thinking makes it so" in Shakespeare's *Hamlet,* cf. also Searle 1975/1979, 73 ff.; Gabriel 1983/1991, 8). While Frege's semantic definition of fiction allows for the division of a literary text into fictional and non-fictional parts, his pragmatic approach is unequivocal: in fiction it is pointless to raise the question of truth with respect to any part of the text—regardless of whether it contains historical or geographical names or universal sentences. Fictional works have no cognitive value. On that issue Frege seems to be an emotivist.[5]

In a letter to Russell dated 28.12.1902 Frege specifies his view on fiction by leaving out the talk of "pseudo-thoughts" and keeping only the talk of "pseudo-assertions". Indeed, he continues to hold that there are thought-expressing sentences which are neither true, nor false (because they contain proper names or definite descriptions without reference) and that *such* thoughts belong to fiction, but nevertheless, according to him, the thoughts expressed in fictional works are no less

---

[5] At any rate, Frege restricts the notion of *knowledge* to the notion of *truth* (true thought): "When someone comes to know something it is his recognizing a thought to be true" (Frege 1924/1979, 266), "we cannot recognize a property of a thing without at the same time finding the thought *this thing has this property* to be true" (Frege 1918/1984, 354). Probably, Frege would not have denied that a work of fiction *as a whole* (by way of a fictional story) can convey certain 'experience' (a position supported by Gabriel 1983/1991), but, as it seems, he would not have called such individual experience "knowledge".

thoughts than those that are bearers of truth-values. In the cases of both sentences and sub-sentential expressions, the sense is independent of reference and hence the question of whether something is a thought or not cannot depend on the circumstance of whether it is true, false or has no truth-value in our world. That is why the phrase "pseudo-thought" is misleading. Besides, the phrase is incapable of distinguishing fictional discourse in its specificity, as the example with *Don Carlos* demonstrates.

The retaining of only the talk of "pseudo-assertions" seems to testify to a greater weight given to the pragmatic as compared to the semantic account of fictional discourse. Whereas the question of the (pseudo-)thought that *p* concerns the question of the *sense* of the sentence "*p*" (i.e., the question of the truth-value-capable part—or, in other words, of the descriptive core—of "*p*"'s content), the question of (pseudo-)asserting "*p*" arises in the context of the questions concerning what we *do*, when we utter a sentence (assert something, ask about something, command something; or, namely: pretend to assert something, pretend to ask about something, and so on). Now Frege says: thoughts are *expressed* in fiction too. The only difference with the ordinary speech lies in the fact that in fiction they are not *asserted*. "This is also why a poet [Dichter] cannot be accused of lying if he knowingly says something false in his poetry [Dichtung]" (Frege 1902/1980, 152). Thus, Frege shifts the emphasis in the characterization of fictional discourse from the distinction between thoughts being true or false and thoughts being neither true, nor false to another distinction of his: the one between expressing and asserting a thought.

Before we can establish whether a thought is true or not, we have to be able to grasp it. Frege calls the grasping of a thought "thinking" and the acknowledgement of its truth "judging" (1918/1984, 355 ff.; cf. earlier 1892a/1984, 164, fn. 10). These are two mental acts (1897/1979, 145; 1918/1984, 368) with the second one presupposing the first (1915/1979, 251; 1918/1984, 356; 1924/1979, 267). They are connected with certain speech acts, since, according to Frege, we anyway do not have any other access to a thought except by means of sentences (1924/1979, 269). To grasp a thought, we need, therefore, its linguistic expression. In order to convey the judgement that we recognize a thought as true, we also need—apart from the linguistic means for its expression—expressive means for its assertion (1906b/1979, 198). In natural languages, however, the expressive means for these two activities are intertwined (1899/1979, 168; 1906a/1979, 185). Whoever utters a simple assertoric sentence in an ordinary context, e.g.,

(8)   The sun has risen.

makes a truth-claim to what is said. In cases like (8) expressing a thought and asserting it coincide, because the two are achieved through the form of the assertoric sentence (1906b/1979, 198, 1915/1979, 252, 1918/1984, 356). To draw attention to the difference between expressing and asserting, Frege gives examples with conditional and interrogative sentences. In the sentence:

(9)   If the sun has risen the sky is cloudy.

we merely express the thought that the sun has risen without asserting that this is the case. The thing whose truthfulness we are committed to is that the cloudiness of the sky would obtain *if* the condition that the sun has risen is fulfilled (*if* it is true that the sun has risen). In uttering (9), however, we do not say whether that condition is fulfilled. Similarly, in the question:

(10)   Has the sun risen?

the thought of the sunrise is expressed without being asserted as true. With (10) we request an assertion in the sense: "(8)—yes or no?".

In the process of elaborating these distinctions Frege points to fictional discourse as an example of the use of language in which we express thoughts without (properly) asserting them. Expressing and asserting thoughts are not linguistic acts of the same order: expressing, just like designating, is a semantic category by which the relations between linguistic units and the world are described whereas asserting is rather a pragmatic category—it is one of the things we perform by uttering sentences in given speech situations. In that connection Frege introduces the term "assertoric force" (*behauptende Kraft*) by which it is indicated whether in expressing a thought an assertion is made or not. For example, while the words "the sun has risen" are uttered with assertoric force in (8), in (9) they are uttered without assertoric force although the whole complex sentence (9) is uttered with assertoric force. Using Searle's terminology, we can say: assertoric force is one of the illocutionary forces by which speech acts are performed. For example, the speech act performed by (10) has the same propositional content as the one performed by (8)—(10) and (8) express the same thought, as Frege would put it—but has a different illocutionary force, namely, an interrogative one (cf. Searle 1969, § 2.4).

In the thematization of sense-only-sign (i.e., of truth-valueless sentences and reference-less proper names), Frege's example of fiction was speech containing mythological or legendary proper names ("Odysseus", "Nausicaa", "Scylla" and "William Tell"). Then, in the pragmatic account there appeared the example of the proper name "Don Carlos" (that can have both a fictional and non-fictional use). Now, when proceeding from the distinction between expressing a thought and asserting it, the figure of an actor on stage and his speech comes to the foreground.[6] Frege presents things as follows: the actor on stage utters sentences without assertoric force (Frege 1906a/1979, 194; cf. also 1915/1979, 251). And more generally: in fiction one speaks without assertoric force (Frege 1906b/1979, 198).

Let us, to conclude this topic, quote two longer passages that can help us see more clearly the relation between an assertion and a pseudo-assertion. The first one comes from the letter to Dingler from 06/02/1917:

> According to my way of speaking, we think by grasping a thought, we judge by recognizing a thought as true, and we assert by making a judgement known. It is one thing merely to express a thought and another simultaneously to assert it. We can often tell only from the external circumstances which of the two things is being done. What an actor says on the stage has usually the form of an assertoric sentence and would also be understood as an

---

[6] Indeed, the example is used as early as in *SB* in a footnote where Frege (1892a/1984, 163) speaks of "signs intended to have only sense" and defines as such signs the words of an actor on stage and even the actor himself and his acts.

assertion if it was said off-stage; but we know that on stage it is not said in earnest [Ernst], but only playfully [nur Spiel]. The actor only acts *as if* he were asserting something, just as he only acts *as if* he wanted to stab someone, and he cannot be charged with lying any more than with attempted murder. What is spoken on stage is said without assertoric force. But in the language of science, too, a thought is sometimes merely expressed without being put forward as true, e.g., in interrogative sentences and conditional clauses. That is why I distinguish between thoughts and judgements, expressions of thought [Gedankenausdrücke] and assertions. (Frege 1917/1980, 20 [my emphasis and in part my translation].)

Here, the speaking-as-if that we introduced at the beginning of the paper is explicitly taken to be a characteristic feature of fictional discourse (cf. also Frege 1914a/1979, 233). Let us take a look at a passage from "Der Gedanke", where Frege discusses the relation between assertions and as-if-assertions (pseudo-assertions):

An advance in science usually takes place in this way: first a thought is grasped, and thus may be expressed in a yes-no question; after appropriate investigations, this thought is finally recognized to be true. We express acknowledgment of truth in the form of an assertoric sentence. We do not need the word "true" for this. And even when we do use it the properly assertoric force does not lie in it, but in the assertoric sentence-form; and where this form loses its assertoric force the word "true" cannot put it back again. This happens when we are not speaking seriously [im Ernste]. As stage thunder is only pseudo-thunder and a stage fight is only a pseudo-fight, so stage assertion is pseudo-assertion. It is only acting [nur Spiel], only fiction. When playing his part the actor is not asserting anything; nor is he lying, even if he says something of whose falsehood he is convinced. In fiction we have the case of thoughts being expressed without being actually put forward as true, in spite of the assertoric form of the sentence [...]. Therefor the question still arises, even about what is presented in the assertoric sentence-form, whether it really contains an assertion. And this question must be answered in the negative if the required seriousness [Ernst] is lacking. (Frege 1918/1984, 356. In part my translation.)

When Dummett considers the passage and similar ones in Frege's works, he criticizes Frege for presenting things in such a way that it leaves the impression that the actor on stage does something "less" than making assertions (let say, merely expressing thoughts). On the contrary, according to Dummett, the actor does something "more": he is *acting* the making of assertions (Dummett 1973/1981, 311). If we have at our disposal a special assertion-sign, such as the one Frege introduces in his logical notation, and means for a plain expression of thought, we can represent the asserting of the thought that p as follows:

⊢ (p)

where "⊢" indicates the assertoric force, and "(p)" stands for the expression of the thought that p. If we now add a new sign for, let say, 'fictional force', e.g., "🎭", what the actor on stage or the narrator of a fictional story do,[7] when they utter assertoric sentences can be symbolized thus:

---

[7] Although it may seem that while the actor on the stage first and foremost speaks dirctly, the narrator has to use indirect speech, it is, nonetheless, quite possible for an actor to report somebody's words as well as for a narrator to speak about himself (or to let a fictional character tell the story instead of him). The distinction between simple statements and indirect speech is not really relevant here, but whether the words spoken or written should be taken seriously or not.

🎭 ⊢ (p)

Here the "'more'' demanded by Dummett would be expressed by "🎭". "🎭" and "⊢", however, should not be merged in a single sign for 'fictional assertion' because that would double the symbolic presentation of all speech acts, since then we would need, e.g., also a sign for fictional question, fictional command and so on, whereas now by means of the sign for 'fictional force' we would simply present "the acting of asking the question whether *p*" thus:

🎭 ? (p)

The symbolization of the illocutionary act of asking by "? (*p*)" is borrowed from Searle (1969, 31).

Is the form of as-if-assertions adequately represented by "🎭⊢ (*p*)"? Dummett seems to be of the opinion that something is added to a proper assertion which turns the latter into an as-if-assertion or a pseudo-assertion. But is the theatrical thunder or the theatrical battle, given by Frege as comparisons, a real thunder or a real battle with an additional something? Is the drawing of a pipe a real pipe plus an additional fictional element? The drawn pipe and the enacted assertion are not a real pipe and a real assertion, although in the context of the drawing or the theatrical performance they *represent* precisely a pipe and an assertion. At any rate, we know: with the enacted assertion the question of truth is inappropriate.

8.5 Fictional Truths Without Fictional Objects

So far we have considered fictional discourse in the sense of the language game that the narrator of fictional stories, the actor on stage, etc. play. We have seen that what is characteristic of the assertoric sentences that the narrator, actor etc. utter is that by them, in their use in the respective literary or theatrical context, no truth-claim is being made. But could we not *truthfully* state that, for instance,

(11)   Odysseus was the king of Ithaca
or that
(12)   Sherlock Holmes lived on 221B Baker Street
and by this consider the negations of (11) and (12) to be false? Could we not *truthfully* say also that:

(13)   Odysseus did not exist.

Could we not also commit to the *truth* of (1) and (2)? Are all of these sentences not true? (11) and (12), (13) as well as (1) and (2) as if present examples of fictional discourse of which the question of truth is completely justified. And yet: how do we make true assertions of things that obviously do not exist?

As already mentioned, a distinction should be made between speaking *in* fiction (the activity performed by the narrator in a fictional story) and speaking *of* fictional works and characters. Whereas in speaking in fiction we have an activity of speaking-as-if, in which the question of truth is inappropriate, in speaking about fiction we make assertions that could very well be true or false. Frege noticed that distinction (cf. n. 4). In contrast to the attention paid to speaking in fiction, however, he does not discuss speaking about fiction in his works. Nevertheless, by his notion of 'sense' he provides us with an instrument for analysis of that kind of speaking as well.

When in his "Prolegomenon to a Fregean theory of fiction" Künne (1995, 141) considers *true* sentences containing fictional proper names he distinguishes five types of such sentences: (i) intra-fictional sentences like (11) and (12); (ii) inter-fictional sentences such as "Anna Karenina did not live in the same country Madame Bovary lived in" or "Don Juan had an obsession that Don Quixote did not have"; (iii) ontological sentences like (13); (iv) trans-fictional sentences like (1) and (2), which seem to place fictional and real objects in a relation; and (v) sentences which represent mixed cases of (i–iv). Since sentences of type (ii) can be analyzed as a more complex variant of sentences of type (i), and those of type (v) are just various combinations of sentences of the other types, here I will consider only sentences of the types (i), (iii) and (iv).

Are (11) and (12) really true? If we take them literally—according to Frege's account—they are neither true nor false because they contain proper names without reference, and, as said above, the possession of reference by proper names in a sentence is a prerequisite for the truth or falsity of that sentence, and, respectively, for making an assertion by uttering it. The object to which we refer by the proper name is what the predicate is applied to or denied of. If it does not exist, then the predication, it appears, would be impossible. What we talk about—using singular sentences such as (11) and (12)—in ordinary speech are the *references* of singular terms. Frege, however, says that there are cases, e.g., indirect speech, statements of epistemic attitudes etc., in which the *sense* of a singular term becomes its reference. Is this the case with (11) and (12)?

As both apologists and critics of Frege have noticed, it is the idea of the transformation of sense into reference that provides the perspective from which sentences of type (i) are to be analyzed in the spirit of Frege (cf. Gabriel 1970; Parsons 1982; Künne 1995). (11) and (12) are not to be taken literally. They are elliptical sentences expressing thoughts that get full expression, e.g., in:

(11*)   It says in Homer's poems that Odysseus was the king of Ithaca,
(12*)   It says in Conan Doyle's novels that Sherlock Holmes lived on 221B Baker Street.

(11*) and (12*) are true. They can be denied and then we will get false sentences. The words "It says in [the relevant body of literature] that ...", which occur in them introduce an 'indirect context' where words have 'indirect reference' and this is—as Frege says—their customary sense (*gewöhnlicher Sinn*). In Carnap's language the words "It says in [the relevant body of literature] that …" create an intensional context, i.e., a context where the principle of substitution does not hold for extensionally equivalent expressions (co-referential expressions) but only for intensionally equivalent expressions (synonymous expressions). Künne (1995, 145 ff.) interprets these and similar words as a 'narrative operator', which can be symbolized, for example, as: "N_x: …" (where "x" is a placeholder for the name of the relevant fictional work). Thus (11*) and (12*) would be expressed as follows:

(11**) N_1: Odysseus was the king of Ithaca,
(12**) N_2: Sherlock Holmes lived on 221B Baker Street.

Let "Fa" be a sentence containing the singular term "a". If "Fa" appears in the scope of "N_x: …", then the truth of the complex sentence "N_x: Fa" will not depend on the *truth* of "Fa", but on the *thought* that "Fa" expresses (i.e., on the sense of "Fa"). That is why the truth of "N_x: Fa" will not depend on the reference of "a" either (which is irrelevant to the thought that "Fa" expresses), but only on what "a" contributes to the expression of that thought (i.e., only on the sense of "a"). Because the reference of "a" does not affect the thought expressed by "Fa", "N_x: Fa" can be true even when "a" has no reference. Thus, a sentence and its parts 'change' their references in the scope of "N_x: …". The sentences begin to stand for the thoughts they ordinarily express and thereby the singular terms we are interested in now begin to stand for the parts of the thought they ordinarily express.

There are some commentators such as Parsons (1982) who appeal to Frege's idea that what a singular term refers to in an intensional context is its sense, in order to argue that, according to a Fregean theory of fiction, fictional characters such as Odysseus or Sherlock Holmes are, in fact, some specific sort of intensional objects, called "individual concepts" (a term borrowed from Carnap). This, in my opinion, implies that the stories about Odysseus and Sherlock Holmes do not speak of concrete objects—objects that could have spatio-temporal dimensions—but rather tell about events that happened to abstract objects. This seems to explain the fact that Odysseus and Sherlock Holmes cannot be found among the concrete objects in space and time, but, on the other hand, gives rise to difficulty of explaining how abstract objects could have concrete properties such as, e.g., being married (Odysseus) or not being married (Sherlock Holmes)?

What is Frege's position on this? What does he understand by the 'sense' of a singular term? An intensional object? An individual concept? Parsons correctly notices that while Frege has technical terms that describe the references of singular terms ("objects"), predicates ("concepts") and sentences ("truth-values") as well as terms for the sense of sentences ("thoughts"), he does not have terminology for the sense of singular terms and the sense of predicates (cf. in more detail Polimenov 2013, 17). It is also notable that Frege always explains what the sense of a singular term consists in either (a) with regard to the *object* which we refer to by it (i.e., with respect to its reference) or (b) with regard to the *thought* expressed by the respective sentence that contains it.

In context (a) Frege says: sense is a way of determining the reference (*Bestimmungsweise*) or a mode of its presentation (*Art des Gegebenseins*), or a "path" (*Weg*) of arriving at it. If this is so, how can there possibly be singular terms that have sense but no reference, if sense, as it appears, is precisely a relation between the sign and its reference (a relation of determining it, presenting it in a certain way or of arriving at it through a certain 'path')? We can put it that way: sense contains criteria for identifying reference—criteria by means of which we pick out an object among all other objects from the respective domain as the object of which we intend to state a certain predicate. Whoever understands a singular term grasps certain criteria for identifying an object. These criteria, however, might not be satisfied by any real object. In that case we say: the singular term has sense but no reference.

In context (b) Frege explains: in understanding a sentence, we grasp the thought expressed by it. In order to understand the sentence, we need to understand its parts, including the singular term. In understanding a singular term, we grasp what the singular term contributes to the expression of the thought. And that is its sense. Now if we combine contexts (a) and (b), we can say: a singular term contributes to the expression of a thought by providing the criteria for the identification of the object that the thought is about.

If to grasp a thought means to understand the sentence that expresses it, let us see what it could mean to understand a (simple) sentence, e.g.,

(14)   The evening star is a sunlit celestial body.

Does understanding (14) mean knowing its *truth condition*? (14) is true if the *object* which we refer to by "the evening star", i.e., the planet Venus has the property of being a sunlit celestial body. It is the same truth condition that the sentence:

(15)   The morning star is a sunlit celestial body

has, since the *object* which we refer to by "the morning star" is the *same* as the one which we refer to by "the evening star". Frege, however, insists: (14) and (15) express *different* thoughts because it is possible for someone who understands both (14) and (15) to reckon one of the sentences true and the other one false. What the thoughts expressed by (14) and (15) differ in are the different criteria for object identification that they contain. It is possible and even quite probable that in understanding (14) and (15) and in grasping in their sense two *different* criteria for the identification of the objects which the thoughts expressed by (14) and (15) are about, we do not know that these criteria are satisfied by the same object. A person who could find himself in this situation would say, for example: I know that the star in the evening sky that is seen then and there is in fact not a star but a planet, though I have no idea whether the star in the morning sky that is seen then and there is a planet too. Whereas the *truth-value* of a sentence depends on the *references* of the singular term and the predicate that occur in that sentence (and we describe that dependence by indicating the *truth conditions* of that sentence) the grasping of the thought that a sentence expresses involves something more than knowing the truth conditions because it is precisely the modes of identification or presentation of the respective objects and concepts that are constitutive for the thought. While the 'truth' of "Fa" depends on whether the *object a—no matter how we refer to it*—has the property of being F, the 'thought' that "Fa" expresses is constituted by the occasion that the object *identifiable as a—whichever it is*—has the property of being F.

We can best elucidate Frege's opinions on the references of sub-sentential expressions from the point of view of the truth of the whole sentence. *The references of the singular term and the predicate are what the truth of a sentence depends on.* In ordinary speech these are the objects which we designate by the singular term and the principles of classification that the predicate stands for. In a complex sentence, however, which contains a subordinate clause in such a way that the truth of the whole complex sentence depends on the *thought* but not on the truth-value of the clause, the *senses* and not the references of the singular term and the predicate of the clause would be what the truth of the whole complex sentence would depend on. It is in this manner that those senses would play the role of the 'references' of the singular term and the predicate in the sentence. Let us take an example: if we symbolize the words "x believes that ..." by "B_x: ..." we could form the following complex sentence:

(16) B_x: the evening star is a sunlit celestial body.

In (16) we cannot substitute "the evening star" by "the morning star" *salva veritate*, because that would change the thought expressed by the subordinate clause, which could change the truth-value of (16). If we assume that "the evening star" has the same sense as the expression "the star in the evening sky that is seen then and then, there and there" we could now put this expression in the place of "the evening star" in (16) *salva veritate*. Therefore, the truth-value of (16) depends on the sense and not on the reference of "the evening star". It is the mode of identification that we grasp when we understand "the evening star" that is constitutive for the truth-value of (16). That is why this mode of identification or presentation which ordinarily constitutes the sense of "the evening star", as in (14), is in (16) the reference of "the evening star".

Is there a place here for some intensional object or individual concept? Can we say that while in (14) by "the evening star" we refer to a celestial body, in (16) by this expression we refer to an intensional object? Is it of an intensional object that we say in (16) that someone believes that it is a sunlit celestial body? If this were so, then the person in question could express in direct speech the thought we ascribe to him/her to believe in by saying: "The mode of presentation of 'the evening star' is a sunlit celestial body" or "The sense of 'the evening star' is a sunlit celestial body". All that seems quite unconvincing since by it we would accept that it is possible for an abstract object to have a property, which, it appears, we can ascribe in a manner that makes sense only to objects in space and time.

Frege accounts for belief sentences of the sort of (16) as follows: in them persons and thoughts are placed in relation. That is why in (16) the thought that the evening star is a sunlit celestial body is not *expressed* but *designated* (cf. Frege 1904/1980, 165). *Mutatis mutandis,* we could say of our examples of intra-fictional sentences: the thoughts that are *expressed* in (11) and (12) are *designated* in (11**) and (12**), but there they are only part of the thoughts that (11**) and (12**) express themselves. Frege, as a footnote of the paper draft "Logic" suggests (cf., n. 4), would have interpreted the conveyance of (parts of) the content of a fictional work in terms of thoughts and indirect contexts. Sentences that we use to convey such contents state relations between fictional works and thoughts. Therefore, if we are to explain 'intra-fictional' speaking in a Fregean manner, we will not need, as it seems, to assume that fictional characters are kind of intensional objects.

Let us now consider ontological sentences of type (iii) as well as a few similar ones. Let us compare the true statements:

(17) Sherlock Holmes is a fictional character

and

(18) Sherlock Holmes is not an historical person.

(17) and (18) do not allow the move of adding the narrative operator. We cannot say: "N_2: Sherlock Holmes is a fictional character" or "N_2: Sherlock Holmes is not an historical person" simply because those things are not said in Conan Doyle's novels. If we use a distinction of Wittgenstein's, we can say: that Sherlock Holmes is a fictional character is not *said* but is *shown* in Conan Doyle's novels. Whoever understands the way Conan Doyle speaks in his novels and by this understands also that "Sherlock Holmes", as used in them, only behaves *as if* it names a certain person, already knows that (17) is true in some sense. Since in (17), however, the name "Sherlock Holmes" is 'used' and not 'mentioned', (17) is not entirely correct from a semantic point of view because the 'use' of the name presupposes that it has a reference, which means that an object (the name's referent) exists in a certain sense, while in another sense this very same object does not exist (namely, in the real spatio-temporal world). That is why (17) can be re-formulated meta-linguistically:

(17*) "Sherlock Holmes" is a fictional proper name.

Frege (1906a/1979, 191) makes a similar move when he says that the correct expression of the thought expressed in (13) is:

(13*) "Odysseus" designates nothing.[8]

(13*) is also the way in which Frege paraphrases sentences such as (18). What matters in this case is that in this way speaking of fictional objects is reduced to speaking of fictional names and fictional works.[9]

The hardest to explain are trans-fictional sentences. With them the move of adding the narrative operator is not possible either. We cannot say, e.g., instead of (1): "It says in Conan Doyle's novels that Sherlock Holmes is a fictional detective who is more famous than any real detective, living or dead", because (1) says things that are not part of the content of the relevant fictional work but, as it seems, facts from the real world.

Thinking that he defends a Fregean theory of meaning, Church (1956, 8) declares that the sentence "Schliemann sought Troy" does not assert a relation between Schliemann and Troy (since at the time Troy was considered to be just a fictional

---

[8] Cf. also the metalinguistic way in which Frege (1892a/1984, 168) says that Kepler did (not) exist.
[9] Künne (151 ff.) suggests replacing the metalinguistic "$\neg \exists x$ ('Odysseus' designates x)", where "Odysseus" is mentioned, by the narrative operator and rendering the thought as: "$\neg \exists x$ (N_1: Odysseus = x)", where "Odysseus" is used.

city) but rather a relation between Schliemann and the concept of Troy.[10] However, we cannot really impute to Schliemann that he was seeking a concept (a fictional object). At the very least, if he had decided to "seek" a concept he would hardly seek it in space and time. Frege would have said that the moment he started looking for Troy (or its remains) Schliemann left fiction and entered the realm of science. By that, however, he had already presupposed that the proper name "Troy" had a reference, although he himself did not know its precise location.

But what was Homer doing when he was composing the *Iliad*? Was he only behaving *as if* he was making assertions, or, on the contrary, believed that he was describing 'historical' persons and events (depicted, of course, from the perspective of the then natural mythological dimensions of the world)? Since this question would lead us away from the topic at hand, let us instead ask—as the intentions of, e.g., Conan Doyle seem far more clear to the contemporary reader—whether it would make sense for anyone to start looking for Sherlock Holmes (or traces of him) in the real world? From the point of view of Frege's conception of fictional discourse that would be completely meaningless and an indication of the fact that this person does not understand the language game Conan Doyle plays in his novels. But Lewis (1978, 39) points out that it is still *possible*, though very unlikely, on the one hand, that Conan Doyle wrote his Sherlock Holmes stories as pure fiction and nevertheless, on the other hand, that, by coincidence and without Conan Doyle knowing, there existed in our world a person to whom all the things described in those stories happened and who was even called "Sherlock Holmes".

A similar possibility is discussed by Ryle and Moore in connection to Dickens's *Pickwick Papers* by supposing the existence of a real Mr. Pickwick, of whom Dickens knew nothing. While Ryle (1933, 39) is of the opinion that in this case the sentences of *The Pickwick Papers* would be true of the real Mr. Pickwick (which implies that the name "Mr Pickwick", as it is used in the novel, would refer to the real person), Moore (1933, 69) objects that if Dickens's intention was not to tell the story of the real Mr. Pickwick, then neither the sentences of the novel, as *used* by Dickens, could be about the real Mr. Pickwick, nor the name that occurs in them could refer to him. Frege's position is close to Moore's. The sentences of a novel in so far as only quasi-assertions are made by them could not be about a real person, and hence the names they contain only behave as if they designate someone. That does not, however, preclude the *use* of the same sentences outside fiction to make assertions about a certain person. According to Frege's semantics, someone could leave the realm of fiction and start looking for, say, Sherlock Holmes (more precisely: could try to check whether the name "Sherlock Holmes" is empty) if for some reason he finds, e.g., researching whether Conan Doyle used a prototype etc. interesting. And, conversely, someone could leave the realm of science and let a historical treatise be enacted by actors on stage. The important thing is not to con-

---

[10] Church (1951, 111) calls this type of concepts "individual concepts", giving the following examples: "I am thinking of Pegasus", "Ponce de Leon searched for the fountain of youth" and "Barbara Villiers was less chaste than Diana". These sentences contain the singular terms "Pegasus", "the fountain of youth" and "Diana", that, according to him, designate individual concepts.

fuse these two realms and to remember that in science we are guided by questions of truth, which is irrelevant to fiction.[11]

What happens, however, with sentences (1) and (2)? In their criticism of Parsons (1982), Gabriel (1987/1993, 369–374) and Künne (1995, 156–160) draw attention to the fact that the predicate "famous" is an epistemic one. An object in itself does not have the property of being famous, it is famous only with regard to other objects that *know* it. Can we replace "famous" in (1) and (2) with a predicate that does not include in its content a relation to the intentional acts of other objects? Can we say, e.g.: "Sherlock Holmes (is a fictional detective who) solved more complicated crimes than any other real detective (living or deceased)"? This statement sounds strange, because we know that the fictional character Sherlock Holmes could not possibly solve any real crime. The famousness of a person can be interpreted, as it seems, as a famousness of his/her name. That is why Gabriel (1987/1993, 373) proposes that we paraphrase (1) and (2) thus:

(1*) "Sherlock Holmes" is a fictional detective-name, which is more famous than the name of any real detective, living or dead.
(2*) A certain fictional detective-name is more famous than the name of any real detective.

The way in which Frege specifies ontological sentences of type (iii) suggests that probably he would have considered (1*) and (2*) as a more precise expression of the thoughts of (1) and (2). In any case, it appears that the sentences used to make true assertions about fictional objects can be reduced to sentences about fictional names and works of fiction.

References

Aschenbrenner, K. (1968). Implications of Frege's philosophy of language for literature. *The British Journal of Aesthetics, 8*, 319–334.
Church, A. (1951). The need for abstract entities in semantic analysis. *Proceedings of the American Academy of Arts and Sciences, 80*, 100–112.
Church, A. (1956). *Introduction to mathematical logic: Vol. 1*. Princeton: Princeton University Press.
Dummett, M. (1973/1981). *Frege: Philosophy of language* (2nd ed.). Cambridge, MA: Harvard University Press.
Frege, G. (1980). *Philosophical and mathematical correspondence*. G. Gabriel, H. Hermes, F. Kambartel, C. Thiel, & A. Veraart (Eds.), B. McGuinness (Ed.) of the English Trans., H. Kaal (Trans.). Oxford: Blackwell.
Frege, G. (1884/1950). *The foundations of arithmetic: A logico-mathematical inquiry into the concept of number*. J. L. Austin (Trans.). Oxford: Blackwell.

---

[11] The fact that Frege was a contemporary to Schliemann's discovery is worth noting. We can suppose that the examples with the name "Odysseus" or the sentence "Priam's palace was wooden" in the *Begriffsschrift* (§ 2) were topical for him and his readers.

Frege, G. (1891/1980). Letter to Husserl (24.05). In G. Frege, *Philosophical and mathematical correspondence* (pp. 61–64). G. Gabriel, H. Hermes, F. Kambartel, C. Thiel, & A. Veraart (Eds.), H. Kaal (Trans.). Oxford: Blackwell.

Frege, G. (1892a/1984). On sense and meaning. In G. Frege, *Collected papers on mathematics, logic, and philosophy* (pp. 157–177). B. McGuiness (Ed.), M. Black, et al. (Trans.). Oxford: Basil Blackwell. [*SB*]

Frege, G. (1892b/1979). [Comments on sense and meaning]. In G. Frege, *Posthumous writings* (pp. 118–125). H. Hermes, F. Kambartel, & F. Kaulback (Eds.), P. Long & R. White (Trans.). Oxford: Basil Blackwell.

Frege, G. (1897/1979). Logic. In G. Frege, *Posthumous writings* (pp. 126–151). H. Hermes, F. Kambartel, & F. Kaulback (Eds.), P. Long & R. White (Trans.). Oxford: Basil Blackwell.

Frege, G. (1899/1979). On Euclidian geometry. In G. Frege, *Posthumous writings* (pp. 167–169). H. Hermes, F. Kambartel, & F. Kaulback (Eds.), P. Long & R. White (Trans.). Oxford: Basil Blackwell.

Frege, G. (1902/1980). Letter to Russell (18.12). In G. Frege, *Philosophical and mathematical correspondence* (pp. 152–154). G. Gabriel, H. Hermes, F. Kambartel, C. Thiel, & A. Veraart (Eds.), H. Kaal (Trans.). Oxford: Blackwell.

Frege, G. (1904/1980). Letter to Russell (13.11). In G. Frege, *Philosophical and mathematical correspondence* (pp. 160–166). G. Gabriel, H. Hermes, F. Kambartel, C. Thiel, & A. Veraart (Eds.), H. Kaal (Trans.). Oxford: Blackwell.

Frege, G. (1906a/1979). Introduction to logic. In G. Frege, *Posthumous writings* (pp. 185–196). H. Hermes, F. Kambartel, & F. Kaulback (Eds.), P. Long & R. White (Trans.). Oxford: Basil Blackwell.

Frege, G. (1906b/1979). A brief survey of my logical doctrines. In G. Frege, *Posthumous writings* (pp. 197–202). H. Hermes, F. Kambartel, & F. Kaulback (Eds.), P. Long & R. White (Trans.). Oxford: Basil Blackwell.

Frege, G. (1914a/1979). Logic in mathematics. In G. Frege, *Posthumous writings* (pp. 203–250). H. Hermes, F. Kambartel, & F. Kaulback (Eds.), P. Long & R. White (Trans.). Oxford: Basil Blackwell.

Frege, G. (1914b/1980). Letter to Jourdain (undated). In G. Frege, *Philosophical and mathematical correspondence* (pp. 78–80). G. Gabriel, H. Hermes, F. Kambartel, C. Thiel, & A. Veraart (Eds.), H. Kaal (Trans.). Oxford: Blackwell.

Frege, G. (1915/1979). My basic logical insights. In G. Frege, *Posthumous writings* (pp. 251–252). H. Hermes, F. Kambartel, & F. Kaulback (Eds.), P. Long & R. White (Trans.). Oxford: Basil Blackwell.

Frege, G. (1979). *Posthumous writings*. H. Hermes, F. Kambartel, & F. Kaulback (Eds.), B. McGuinness (Ed.) of the English Trans. Oxford: Blackwell.

Frege, G. (1917/1980). Letter to Dingler (06.02). In G. Frege, *Philosophical and mathematical correspondence* (pp. 19–23). G. Gabriel, H. Hermes, F. Kambartel, C. Thiel, & A. Veraart (Eds.), H. Kaal (Trans.). Oxford: Blackwell.

Frege, G. (1918/1984). Logical investigations. Part I: Thoughts. In G. Frege, *Collected papers on mathematics, logic, and philosophy* (pp. 351–372). B. McGuiness (Ed.), M. Black, et al. (Trans.). Oxford: Basil Blackwell.

Frege, G. (1919a/1984). Logical investigations. Part II: Negation. In G. Frege, *Collected papers on mathematics, logic, and philosophy* (pp. 373–389). B. McGuiness (Ed.), M. Black, et al. (Trans.). Oxford: Basil Blackwell.

Frege, G. (1919b/1979). [Notes for Ludwig Darmstaedter]. In G. Frege, *Posthumous writings* (pp. 253–257). H. Hermes, F. Kambartel, F. Kaulback (Eds.), P. Long & R. White (Trans.). Oxford: Basil Blackwell.

Frege, G. (1923/1984). Logical investigations. Part III: Compound thoughts. In G. Frege, *Collected papers on mathematics, logic, and philosophy* (pp. 390–406). B. McGuiness (Ed.), M. Black, et al. (Trans.). Oxford: Basil Blackwell.

Frege, G. (1924/1979). Sources of knowledge of mathematics and the mathematical natural sciences. In G. Frege, *Posthumous writings* (pp. 267–274). H. Hermes, F. Kambartel, F. Kaulback (Eds.), P. Long & R. White (Trans.). Oxford: Basil Blackwell.

Frege, G. (1984). *Collected papers on mathematics, logic, and philosophy*. B. McGuiness (Ed.). Oxford: Basil Blackwell.

Gabriel, G. (1970). Frege über semantische Eigenschaften der Dichtung. *Linguistische Berichte, 8*, 10–17.

Gabriel, G. (1983/1991). Über Bedeutung in der Literatur: Zur Möglichkeit ästhetischer Erkenntnis. In G. Gabriel (Ed.), *Zwischen Logik und Literatur: Erkenntnisformen von Dichtung, Philosophie und Wissenschaft* (pp. 2–18). Stuttgart: Metzler.

Gabriel, G. (1987/1993). Fictional objects? A "Fregean" response to Terence Parsons. *Modern Logic, 3*, 367–375. Trans. of Gabriel, G. (1987). "Sachen gibt's, die gibt's gar nicht." Sind literarische Figuren fiktive Gegenstände? *Zeitschrift für Semiotik, 9*, 67–76.

Gabriel, G. (1991). Der Logiker als Metaphoriker: Freges philosophische Rhetorik. In G. Gabriel (Ed.), *Zwischen Logik und Literatur: Erkenntnisformen von Dichtung, Philosophie und Wissenschaft* (pp. 65–88). Stuttgart: Metzler.

Künne, W. (1995). Fiktion ohne fiktive Gegenstände: Prolegomenon zu einer Fregeanischen Theorie der Fiktion. In J. L. Brandl, A. Hieke, & P. M. Simons (Eds.), *Metaphysik: Neue Zugänge zu alten Fragen* (pp. 141–161). Bonn: Akademia Verlag.

Lewis, D. (1978). Truth in fiction. *American Philosophical Quarterly, 15*, 37–46.

Moore, G. E. (1933). Imaginary objects. *Proceedings of the Aristotelian Society, Supplementary Volume, 12*, 55–70.

Parsons, T. (1975). A Meinongian analysis of fictional objects. *Grazer Philosophische Studien, 1*, 73–86.

Parsons, T. (1980). *Nonexistent objects*. New Haven: Yale University Press.

Parsons, T. (1982). Fregean theories of fictional objects. *Topoi, 1*, 81–87.

Polimenov, T. (2013). Teile des Sinns und Teile der Bedeutung. *Wismarer Frege-Reihe, 2*, 11–31.

Ryle, G. (1933). Imaginary objects. *Proceedings of the Aristotelian Society, Supplementary Volume, 12*, 18–43.

Searle, J. (1969). *Speech acts: An essay in the philosophy of language*. Cambridge: Cambridge University Press.

Searle, J. (1975/1979). The logical status of fictional discourse. In J. Searle (Ed.), *Expression and meaning: Studies in the theory of speech acts* (pp. 58–75). Cambridge: Cambridge University Press.

Textor, M. (2011). Sense-only-signs: Frege on fictional proper names. *Grazer Philosophische Studien, 82*, 375–400.

Chapter 9
The Dichtung of Analytic Philosophy: Wittgenstein's Legacy from Frege and Its Consequences

Allan Janik

Abstract Wittgenstein's attitude to writing philosophy is an important part of his complex legacy from Frege. Even the frequently misconstrued phrase, "Philosophie dürfte man eigentlich nur dichten", is part of that legacy. How should we actually render that sentence in English? How is the idea that *Dichtung* is a necessary aspect of philosophical method rooted in thoughts that ultimately find their way back to Frege? Where do we find *Dichtung* in the so-called private language argument? How is Wittgenstein's view of the role of *Dichtung* in philosophy related to his appreciation of humor in philosophizing? What is the role of humor in Frege's work? How does Wilhelm Busch in *Eduards Traum* illustrate the philosophical significance of humorous *Dichtung*? What is the link between *Dichtung* and craftsmanship in writing philosophy for Wittgenstein? These are the central issues that the article addresses.

Keywords Examples • Fiction in philosophy • Frege • Objectivity • Philosophical practice • Self-knowledge • Busch, Wilhelm • Wittgenstein

Wittgenstein's legacy from Frege is immensely complicated. It is not simply a matter of inherited problems about the nature of propositions, meaning and truth but also of his conception of the strategy for dissolving philosophical problems and the appropriate style that said strategy entailed, which slowly but surely emerged in the course of his whole career. It too was profoundly influenced by Frege. We forget that Frege was originally at the top of the figures who influenced his philosophical method of clarification as he reflected on the matter around 1931 (Wittgenstein

---

A. Janik (✉)
University of Vienna, Vienna, Austria

University of Innsbruck, Innsbruck, Austria
e-mail: allan.janik@uibk.ac.at

1998, 16; the text has no more exact date). In fact he spent his whole life digesting Frege as it were. What follows is meant to present an important example of that.

Let us begin *in medias res*. In the soon 40-odd years since Peter Winch published his first translation of the text known as *Vermischte Bemerkungen* (Wittgenstein 1980, 1998) the importance of the sentence "philosophy ought really to be written only as a *poetic composition*" (Wittgenstein 1980, 24e), has become inordinately overemphasized in the literature on Wittgenstein. At the same time it has been read superficially to the detriment of understanding the nature of Wittgenstein's philosophy. Philosophers in the so-called "Continental" tradition and their compatriots in comparative literature have taken it as proof positive that Wittgenstein was anything but the analytic philosopher par excellence as he was for the Vienna Circle (with the notable exception of Otto Neurath). Analytical philosophers of the strict observance, for their part, have been all too anxious to acquiesce to this view. For them, the lack of conventional arguments in Wittgenstein's mature works and, most simply, his failure to come to the point in a systematic way encouraged that view. Having absorbed what they could from his Sibylline books they are now prepared to consign him to be the junkheap of "Continental" philosophy. Philosophical poetry is for them, as it was for the Vienna Circle, metaphysics of the most deplorable kind. Moreover, the surprising, even, captivating power of Wittgenstein's imaginary counter-examples along with his lapidary style of writing in all his works have strongly reinforced the image of Wittgenstein the maverick philosopher-poet thumbing his nose at mainstream analytical philosophers. So it is anything but implausible to read Wittgenstein as philosopher-poet maudit. However, a close reading of the text itself along with an even closer examination of his highly ramified, if superficially simple, concept of philosophical methodology[1] leads to a *very* different conclusion. It is undeniable that Wittgenstein saw philosophy as a confrontation with "scientific" thought whose aim is to establish its limits from within. This explains why Wittgenstein's philosophical interlocutors were *always* analytical philosophers.[2] He avoided all contacts with other philosophers. His dialogue was exclusively with them. All of this, in turn, provokes a profound re-evaluation of his actual place in the drama that was twentieth century philosophy, which is completely foreign to, say, post-modern Wittgenstein interpreters (which makes it doubly lamentable that analytical philosophers have played along here).

The place to begin is with Peter Winch's two translations of Wittgenstein's sentence: "[P]hilosophie dürfte man eigentlich nur *dichten*" (Wittgenstein 1998, 28).[3]

---

[1] This was my goal in writing *Assembling Reminders* (Janik 2009).

[2] It is telling that when Wittgenstein visited Vienna in early 1950 he participated in philosophical discussions in the circle around Viktor Kraft and not those in the circle around Ludwig Hänsel, who was one of his closest friends. The action was where the analytical philosophers were, nowhere else.

[3] The remark reads in full: "Ich glaube meine Stellung zur Philosophie dadurch zusammengefaßt zu haben, indem ich sagte: [P]hilosophie dürfte man eigentlich nur *dichten*. Daraus muß sich, scheint mir, ergeben, wie weit mein Denken der Gegenwart, Zukunft, oder der Vergangenheit angehört. Denn ich habe mich damit auch als einen bekannt, der nicht ganz kann was er zu können wünscht." (Wittgenstein 1998, 28).

9 The Dichtung of Analytic Philosophy: Wittgenstein's Legacy from Frege and Its... 145

The first one runs: "philosophy ought really to be written only as a *poetic composition*" (Wittgenstein 1980, 24e). The second reads as follows: "really one should write philosophy only as one *writes a poem*" (Wittgenstein 1998, 28e).

It may seem strange to proceed from a translation rather than the original text. However, much is to be said for the curious fact that this text, like the *Vermischte Bemerkungen* as such, has been principally received internationally as *Culture and Value* in the first of Winch's translations, which is still readily available in bookstores despite being long dated both in terms of it textual basis and, on Winch's own view, the translation.[4] It is the version that most readers seem to have read and reacted upon. Moreover, having been edited from the philosopher's papers by G.H von Wright, in the strict sense it is not even a text by Wittgenstein even though Wittgenstein wrote every word of it.

Anyway, the two translations of Wittgenstein's German sentence do not differ in their problematic elements. Both English versions ascribe a certain striving to write philosophy as poetry to Wittgenstein. This is misleading. There is only a weak and unenlightening analogy between Wittgenstein's philosophical writings and anything that can respectably be described as poetry. Furthermore, if Wittgenstein had wanted to compare his philosophical texts to poetry as conventionally understood, he would have made reference to "Lyrik" in German, which is the normal word for poetry as it is understood in English. "Dichtung" and its verb "dichten"[5] is a more generic term, closer to the English "fiction" of which there are several species. In the classical manuals of grammar "Dichtung" certainly refers to "poetry" in the broad sense including the species epic, dramatic and lyric poetry.[6] In her critique of Winch, Marjorie Perloff (2011, 714f.) has gone so far as to say that there is not even an explicit reference to writing in the text, which seems to be stretching a point but is certainly correct in the strict sense. Since Wittgenstein did in fact write philosophy, rather than sing (or whistle!) his philosophizing, a more accurate translation would seem to be something like: philosophy should only really be conceived as fiction (i.e., as a work of the imagination). This is not perfect either but it conforms to Wittgenstein's practice, as we shall see.

The problems with the text do not end there. To begin with it is not entirely clear what Wittgenstein means by the sentence.[7] When Wittgenstein is speaking of

---

[4] In the context of the research project "Wittgenstein's *Culture and Value*: An Electronic Edition" sponsored by the Austrian Science Fund that ran from 2007 and 2010 Kerstin Mayr, Joseph Wang and I carefully compared both translations with Professor Pichler's definitive German text line by line, words for word. We also investigated the text genesis and the initial reception of the work carefully. For reasons beyond our control the results could not be published. They are available at (not from!) the Brenner Archives Research Institute at the University of Innsbruck and the Wittgenstein Archives at the University of Bergen in Norway.

[5] The word is ultimately derived not from *dicht* (thick) but from the Latin *dictare* meaning "to dictate, to set up or to draft" (as one drafts, say, a letter).

[6] This way of dividing up the genus "literature" originating in the eighteenth century is ultimately derived from Friedrich Schlegel. See Jäger (1970, 2–30).

[7] Severin Schroeder made this point to me forcefully and enlighteningly at a Wittgenstein symposium organized by the Bergen University Department of Philosophy in Marifjøra, Norway in 2011.

philosophy, does he mean his own or the philosophical enterprise as such. "Dürfte eigentlich nur" indicates that philosophy is not done this way which turns out to be the only legitimate way to philosophize. Somehow we do not manage to philosophize that way. However, for that very reason, some commentators including the astute Professor Perloff have suggested that the connection with poetry is despite all that appropriate. In a sense this is all well and good for her because her principal concern is with poetry and there are certainly aspects of Wittgenstein's way of writing philosophy that deserve to be termed poetic in the usual sense. She supports her argument by reference to Wittgenstein's own complaint that, despite all his considerable efforts, he cannot do what he wants to in philosophy (cf. Wittgenstein 1958, Preface). On that basis, she suggests that he is really aiming at writing poetically. However, that view conveniently overlooks the fact that what he actually said is that he cannot *quite* do what he wants to do. I take that to mean, on the contrary, that he would like to be yet more convincing in his own way of doing philosophy not that he would scrap his prose modus operandi if he could somehow be a better philosophical poet. In fact, nothing about Wittgenstein's way of doing philosophy supports the idea that he saw himself as an incompetent primitive hopelessly failing to be a philosophical Rilke. It is much more a matter of perfectionist self-critique, i.e., refining the ways of expressing himself that were essential to his "work of clarification" which he developed in the course of some 25 years by the time that he came to write this, such that his "arguments" would be yet more convincing to analytic philosophers of the strict observance than they are in the form that he has left them. The problem is that Wittgenstein was first last and always a philosopher, even if, as Perloff correctly insists, the "poetic"—on my view the *fictional*—dimension of his undertaking was the key to doing philosophy *rightly* for him. So our question has to be: why must that be so for Wittgenstein?

Like nearly everything about Wittgenstein the reasons for his thinking take us far afield from anything resembling a simple answer. To understand why, we have to begin with his most important mentor in philosophy, Gottlob Frege.[8] It was he, after all, who gave Wittgenstein the idea that his philosophizing was "artistic" (Frege 1989, 19) in their correspondence on the *Tractatus*. Frege certainly intended that as just about the most devastating critique that could be raised against a putative work of philosophy. However, Wittgenstein came to see his work as somehow squaring the circle by being both literary and strictly philosophical at the same time as he would famously write to Ludwig Ficker in 1919 (Ficker 2014, 42).[9] How did someone who considered himself to be profoundly influenced by Frege come to see his work that way? How would precisely this dimension of his philosophizing remain central to him over the radical changes that his work of clarification took in the course of the next 30 years?

We have to begin with the very conception behind Frege's *Begriffsschrift* (1879), which is arguably the most important work in the history of logic. In this work Frege

---

[8] Alois Pichler gave me the first inkling that Frege was involved here at the same Marifjøra symposium.

[9] Walter Methlagl was the first to realize this crucial point.

was able to *show* quite literally that functions were the basis of formal logic. He represents functions as standing in a clear and distinct relation to one another which *shows itself* in the logical language and thereby eliminates all possible ambiguity of the sort that inevitably arises within natural languages. On the basis of these relationships we are capable of producing propositions that can represent the world with absolute precision. Thus Frege's position entails an absolutely strict criterion for separating fact from fiction, *Wissenschaft* from *Dichtung*. There is no room for narrative, let alone story-telling or versifying here.

Yet, certain points relating to Frege are relevant for the train of thought that leads to Wittgenstein's notion that fiction is absolutely essential in philosophy. The first is that Frege (Frege 1893) clearly recognized that even in an axiomatic system some things connected with the use of symbols could only be shown rather than said. Here he makes reference to the importance of the *Wink*, "hint", "gesturing by waving the hand" that shows the perplexed student of logic how to go on where no discursive explanation is available. We shall turn to a further relevant point concerning the Fregean background to Wittgenstein's idea that fiction is essential to philosophy at a later point in our exposé.

To understand what was at stake for Frege here, we need to consider his idea of *objectivity*—a theme that is not entirely uncontroversial within Frege scholarship. In the *Foundations of Arithmetic* he writes that "objective" means: what conforms to law, the conceptual, what can be judged, what can be expressed in words (*was sich ausdrücken läßt*). Frege does not equate what is objective with what is real in the sense that it is given in pure sensory intuition. Frege insists that objectivity is bound to the logical (transcendental?) conditions that make thinking what is true possible in the first place. So the earth's axis and the equator, for example, are objective for him without being real (*wirklich*). Such objective unrealities form the basis of the system of presuppositions upon which scientific knowledge rests. For all their unreality, they are anything but fictional, distinctly not fictions as, say, Frege's contemporary Hans Vaihinger would claim, but objective in an unusual sense that bears significant similarities with the objective unrealities of Plato or even Immanuel Kant. The point is that Frege's realism is considerably less medieval than, say, that of Bernhard Bolzano (at least in the version we find in *The Foundations of Arithmetic*).[10] Wittgenstein shared in principle his preoccupation with this sort of objectivity in a way that profoundly influenced his concept of philosophical method (and probably the course of his life as well).[11]

The Foundations of Arithmetic (1884), the most accessible work by Frege, unlike *Begriffsschrift* (1879) or *The Basic Laws of Arithmetic* (1893, 1903), seems to have been written in the informal mode albeit in scholarly form. There is copious references to the works of other writers explicitly with a view to compensate for the highly unusual, but formally correct presentation of his earlier, neglected,

---

[10] Sluga (1980) argues for a Kantian view of Frege's objectivism against Michael Dummett's more Platonic view of it. However, this Kantian variation is hardly identical with the historical Kant. Sluga's view of Fregean objectivity tends to co-incide with ours as Dummett's does not.

[11] This is spelled out in more detail in the Frege chapter of Janik (2009).

Begriffsschrift in which he first developed the notion of logic based upon the idea of truth "functions" but which was wildly misconstrued. Here Frege goes out of his way to meet *any and all* opinions to the contrary of his own views. In fact, Frege leaves no stone unturned in his efforts to come to grips *comprehensively* with contrary views. Not only the views of Kant and Mill but those of contemporaries like Baumann, Schröder, Schlömilch and nearly every writer on mathematics dissenting from his views is analysed in depth. One informed commentator describes his fourfold approach to the view he would criticize as follows in a way that wholly anticipates Wittgenstein:

> ... he takes remarks by his opponents absolutely literally; he draws consequences with absolute rigor; he brings large-scale claims down to earth with very concrete counterexamples; he searches out hidden inconsistencies. (Sluga 1980, 194)

Frege's critique of, say, Benno Erdmann in the Preface to *The Basic Laws of Arithmetic* (1893, XIVff.), is a perfect example of this strategy, which mirrors Wittgenstein. My point in introducing this aspect of Frege's work here is to remind us that Frege's own corpus is more "literary" than meets the eye and, whether he liked it or not, it *had* to be that way if he was going to be *convincing* to his peers.

The mature Wittgenstein's thought is centered around the notion of following a rule where there are no formal rules but only examples to be imitated. That center dictates that his philosophical method is itself largely a matter of a particular way of employing *examples* where traditional philosophers have had recourse to formal arguments and so-called "theories." It was crucial to Wittgenstein that these examples neither be obvious nor simple. They had to be sufficiently subtle and imaginative to undermine conventional perspectives and put our thinking onto new tracks as it were. Thus Wittgenstein asserts at a crucial juncture in the *Philosophical Investigations* (1958, § 122):

> Our grammar lacks the synoptic perspective [*Übersichtlichkeit*]. A synoptic representation conveys the sort of understanding of precisely the sort that we call "seeing connections". Thus the importance of finding and inventing *intermediate cases*.[12]

Even a cursory glance at the examples that Wittgenstein develops in the *Investigations* indicates that he has a decided preference for invented examples over discovered ones. This is exactly the point where fiction becomes relevant to his philosophical work of clarification as he termed it. Moreover, these fictions, which he terms "language games" often have a decidedly bizarre character. Think only of his famous builders whose only words are "block," "pillar," "slab" and "beam" who we are asked to conceive as "a complete primitive language" (1958, § 2). Although it is possible to interpret Wittgenstein as wanting to show us that words are every bit as much signals to act as representations of something, it is not so easy to conceive this as a *complete* (*vollständig*) language.[13]

---

[12] I have corrected the standard translations of Wittgenstein where required.

[13] I profited from listening to the Norwegian philosopher Harald Johannessen's sobering case against Wittgenstein in Bergen several years ago.

The fictions that Wittgenstein calls language games are supposed to clarify basic confusions that may stem from a one-sided view of language as representation that tempts the theoretically-oriented thinker, i.e., typically the analytical philosopher or anybody who conceives of language as something abstract. This seems to be the sort of error that Kant termed a "transcendental illusion," i.e., a source of profound confusion that is entailed by the philosopher's way of framing questions. However, unlike Kant (1966, 728–730), Wittgenstein does not simply want to analyze the inadequacies of that framework, he wants to *dismantle it from within*. That means restructuring the philosophical imagination that such questioning presupposes. This is what he means when he says that philosophy is a matter of working on yourself; for nobody is immune to such temptations. Philosophy is a work that involves liberating the imagination—and that means your own as well as that of others. That is why fiction is so important for the philosophical enterprise.

This is, of course, all well and good to assert. How should we corroborate such a claim? Here, again, there is a certain temptation to look in the wrong place. The answer is less to be found in Wittgenstein's assertions about philosophy than it is in *his way of practicing it*. So *Culture and Value* or even the so-called "philosophy chapter" in Part I of the *Investigations* are not the place to begin even if they cannot ultimately be ignored. It was a great service of Eike von Savigny continually to remind the community of Wittgenstein scholars that the only part of the *Nachlass* that Wittgenstein himself considered ready for the press was Part I of the *Investigations*, whose philosophical center has traditionally been taken to be the famous "private language argument". This is clearly a text in its final form as Wittgenstein would have presented it to the public. It is at the same time more intimately linked to Wittgenstein's impact upon twentieth century thought than any other. That is what it is of paramount importance here. Our question concerns less what Wittgenstein wanted to say there as what he wanted *to do* to philosophers who might be tempted into thinking that there was such a thing as a private language in the first place.

The first thing to be said here is that the "private language argument" is *not an argument* as that term has traditionally been understood by analytical philosophers from Aristotle and Leibniz to Russell and Carnap (just as the "picture theory" in the *Tractatus* is not a *theory* at all—*nothing* in Wittgenstein fits the conventional categories of academic philosophy). It does not consist of premises and a conclusion but of reflections upon a theme. Its aim is nevertheless a kind of analysis (*Hinterfragen*) of what it would mean for *me* to develop a language to describe *my* innermost personal experiences and perceptions. The basis of Wittgenstein's method is the idea that the philosopher's *imagination* has to be re-trained (a kind of counter-*Abrichtung* you might say). More than 20 times in the course of 84-odd paragraphs that make up the discussion of pain and private language Wittgenstein tries to re-focus the way we think and talk about matters of pain and privacy, i.e., the way we imagine our "interior life". He wants to show us that we really do not know what we mean if we assert that there is such a thing as a private language in which I can describe my personal experience of pain. The instruments with which he proposes to dissuade us are:

penetrating questions (1958, §§ 246, 249)
thought experiments (§ 284)
examples from natural history (§ 250)
comparisons (§ 248)
aphorisms (§§ 255, 297)
commands to look at something in a particular way (§§ 279, 245, 300)
allusive references to usage, the complexity of contexts, what it is appropriate to say in a given situation (*passim*)
polar oppositions, differences, canonical examples ("paradigms") and conceptual boundaries (§§ 272, 268).

Thus he asks us:

...we could imagine... (§ 243)
Could one teach a dog to stimulate pain? (§ 250)
..why do we say, "I can't imagine the opposite?" (§ 251)
What would it be like if humans showed no outward signs of pain? (§ 257)
Let us imagine the following case... (§ 258)
It might be said: ... (§ 262)
"But can (I inwardly) undertake to call THIS 'pain' in the future?" (§ 263)
Let us imagine a table (something like a dictionary) that only exists in our imagination: (§ 265)
Why can't my left hand give my right hand money? (§ 268)
Let us now imagine a use for the entry of the sign "S" in my diary. (§ 270)
"Imagine a person whose memory could not retain what the word 'pain' meant..." (§ 271)
Imagine someone saying: ... (§ 279)
Someone paints a picture in order to show how he imagines a theatre scene. (§ 280)
What gives *us so much as the idea that* living beings, things, can feel? (§ 283)
Couldn't I imagine having frightful pains and turning to stone while they lasted? ... And if that happened in what sense will *the stone* have the pains? (§ 283)

This selection comes from the first 40 paragraphs that make up the discussion. It would be tedious to continue through the rest. That Wittgenstein wants to re-orient the analytical philosopher's imagination seems beyond doubt. The point is that we find *nothing* that even remotely looks like a Goethe or Rilke poem there.

Creating imaginary counter-scenarios like the episode with the builders is thus essential to his way of doing philosophy. Although Wittgenstein insisted that he was not doing natural history, natural history plays a crucial role in his elaboration of the nature of human thinking. Abstract language tempts us to see thought and language as having "a life of their own" as it were apart from human action/practice/life. Thus he came to conceive his task as one of "assembling reminders" (1958, § 127) of how concept formation is rooted in nature. In doing so he was careful to emphasize that developing that view he was not advocating a "scientific" or naturalistic view of knowledge but simply "assembling reminders" of the rootedness of thinking *in the life of the kind of animal that uses language*. He was calling our attention to a number of "general facts" of how human beings develop "mentally" that we simply overlook when doing "science." (1958, II, xii, 230) In themselves these facts are commonplaces but their "simplicity and familiarity" (1958, § 129—Heidegger would say their "everydayness") inclines everybody who thinks abstractly to overlook them as trivial when contemplating the nature of thought and reality.

In Part I of the *Philosophical Investigations* Wittgenstein reminds us that "commanding, questioning, story-telling, chatting are as much a part of our natural history as walking, eating drinking, playing." (1958, § 25) Later he explains what he is doing as "supplying... remarks on the natural history of human beings...observations which no one has doubted, but which have escaped remark only because they are always before our eyes." (1958, § 145) At a later stage in his development (which we are accustomed to call Part II of the *Philosophical Investigations*) he writes:

> If the formation of concepts can be explained by facts of nature, should we not be interested, not in grammar, but rather in nature which is the basis of grammar?... But our interest does not fall back on upon these possible causes of the formation of concepts, we are not doing natural science; nor yet natural history—since we can also invent natural history for our purposes. (1958, II, xii)

He goes on to explain:

> if anyone believes that certain concepts are the right ones pure and simple...—then let him imagine certain very general facts of nature to be different from what we are used to, and the formation of concepts different from the usual ones will become intelligible to him. (1958, II, xiii)

Here it is fairly clear why Wittgenstein has to have a preference for fictitious examples of language games over ones that are simply discovered. Reflection upon actual language games cannot provide the kind of contrasts that have persuasive, judgment-forming character when he is trying to show us the limits of our accustomed ways of thinking. However much Wittgenstein's task might look like social science, it is first, last and always philosophical. Thus *Wittgenstein's philosophische Dichtung* or, perhaps better his *gedichtete Philosophie* turns upon his challenge to our imagination in demanding that we reflect upon whether a rose has teeth or why a dog can't stimulate pain or why my left hand can't give my right hand money, to use examples that are familiar to every reader of Wittgenstein but usually considered nonetheless obscure for all that. Moreover, it is hardly incidental that these excurses involve something disquietingly *funny*. See that chair, he asks. Good, now translate it into French! (1967, § 547).

The peculiar wit in Wittgenstein's thinking has scarcely been noticed by commentators but it is nevertheless there and it is philosophically relevant, for example, in his challenge to consider what the sentence, "a rose has teeth," might mean. Indeed, he assiduously collected nonsense his whole life long. At his death this collection, which he willed to Rush Rhees, was composed of some 130 items that he had begun saving in 1920 and continue to augment up to the late 40s. Wittgenstein, it seems, resolutely believed that nonsense was anything but insignificant and that concern had a philosophical significance (see McGuinness 1966).

It is worth speculating a bit about the origins of that practice. As we have mentioned the practice in all likelihood goes back to his youth. It is probably safe to suggest that Karl Kraus most likely lent the practice its initial impetus. As the ever insightful J.P. Stern put it with respect to Kraus' development at just the time the young Wittgenstein would most likely have encountered him,

> Gradually, after 1905, it is borne in upon him that language—that is the way that a statement is made—bears within itself *all* the signs he needs to understand the moral and ethical quality of that statement and of him who made it. (Stern 1966, 73f)

It is easy enough to imagine two bright young *fin de siècle* Viennese teenagers like Paul and Ludwig picking up Kraus' penchant for letting nonsense "show itself" and making a kind of game out of it (see McGuinness 2006). One knows, in any case that Kraus was the only "Dichter" on Wittgenstein's list of the people who had influenced him up to 1933–1934. Yet, for all Wittgenstein's well documented admiration for Kraus (which in his case is not incompatible with the same sort of scathing critique to which the philosopher subjected himself), we know almost nothing about *how* Kraus influenced him (moreover, "influence" as Wittgenstein uses it is far from what it is commonly taken to be as I have shown elsewhere).[14] However, the part of the story that can be reconstructed begins with nobody else than Frege. Yes, Frege! It seems that Wittgenstein was as inspired by his humor as much as he was with his stringent thinking about logic.

To return to the theme of Frege's influence upon Wittgenstein with which we began this discussion earlier: the third aspect of Frege's influence upon Wittgenstein's view that fiction is absolutely essential in philosophy is striking but rarely mentioned in the literature on Wittgenstein. It is the fact that Frege (like Boltzmann and Schopenhauer)[15] was a writer of considerable *wit*. Frege's wit is nowhere more in evidence than in *The Foundations of Arithmetic*, which abounds in hilarious ripostes to his superficial adversaries. Thus Frege will speak of the kinds of confusions that arise in philosophy as the sort of gaffs we would make were we to speak of a *blue idea, a salty concept* or a *chewy judgment*. This involves a philosophical error of the sort that Ryle called a *category mistake*. As in Frege (and in Kraus), humor is essentially linked to taking one's opponent by his word and exposing the superficiality and latent inconsistency of his position by a rigorous analysis of his mode of expression. Be that as it may, Wittgenstein, too, would tend to see philosophical problems as arising when we are unable to resist temptation to ask an inappropriate question and he would find the result just as laughable as Frege (and Karl Kraus). Moreover, he would find the sort of questions that traditional philosophers pose "mad" in *On Certainty* (1969, § 467) and thus in a sense funny (1967, § 328). The appropriate response could only be "to think crazier than the philosophers" (1998, 86), which means "to climb down from the barren hights of cleaverness into the green valleys of stupidity" (1998, 86). Such remarks are not simply aperçus shot from the hip but fundamental observations about his philosophical "work of clarification" as he called it (1998, 16). An example of Frege's humor comes at the point where he is discussing Immanuel Kant's notion that we have an intuition corresponding to each number similar to the intuition we have of fingers or points when we use these numbers. This seems reasonable enough with respect to 1, 2 and 3 but, Frege asks, what is the intuition of 37,863 fingers like? If Kant's view of the role of intuition in arithmetic was correct we would have to have an intuition of 135,664, 37,863 and 173,527, when we added the first two numbers together to get the third. Moreover, the correctness of the third number as sum of the first and second would have to be immediately apparent, which it obviously is not (Frege 1893, 30). In rejecting

---

[14] See my Janik (2009) for a detailed account of these themes.

[15] This is also further developed in Janik (2009).

Baumann's view that the concept "one" rests upon a mental intuition of undividedness and demarcation, Frege writes:

> If this were correct, we would have to expect that animals as well could have a certain idea of unity. Whether a dog looking at the moon has even a definite idea of what we designate by the word "one"? Hardly? (Frege 1962, 64; my translation)

Similarly Frege would remark, with respect to the notion that we should term a proposition empirical, as opposed to a-priori, if we need to have made observations to be conscious of its content, then the stories of Baron Münchausen would also be empirical in that sense because their author would have had to make a considerable number of such observations to invent them (Frege 1962, 35). Indeed, there can be little doubt that Frege was a master in the deployment of the philosophical witticism: There are no cheap laughs here but penetrating wit with a philosophical point.

For all that Frege was anything but a *Dichter*. Our initial question about the relationship between philosophy and fiction (see Gabriel 1991) will come a long way closer to being answered if we can find a *Dichter* whom Wittgenstein considered philosophically important. In a letter to Rush Rhees in 1950 at the very end of his life Wittgenstein declared Wilhelm Busch to be just such a *Dichter*: "he has the REAL philosophical urge" (Rothhaupt 2013). Moreover, his esteemed friend Rudolf Koder emphasized that he read literature for two reasons: because he admired stylistic eloquence as found in the Mörike of "Mozart auf der Reise nach Prag" and because he considered it philosophically profound as was the case with Busch's "Eduards Traum".[16] Once more, we do not know much beyond that, but we may consider what is normally taken to be its significance within the context of Busch's oeuvre.

A brief summary of *Eduards Traum* will be useful here.[17] Eduard parts company with his body as a thinking mathematical point in his dream and tumbles into a chaotic world in which he has a sequence of increasingly horrifying adventures. His adventures begin in a land of numbers that falls to pieces of itself. From the kingdom of points, the geometric plane, where things only have breadth, his journey continues into pure three dimensional space. It continues in the land of isolated bodily parts where people are dissolved and only their individual parts subsist. Then follows a description of the world as we know it in which everyone has his body back together but it turns out to be a view into an abyss. It begins with the description of the peasant world of the farmer which begins as an idyll but flips into a nightmare as a farmer's wife is dismembered. In what follows town and country are both filled with confidence men, tricksters, idiots and all sorts of criminals. There is not the slightest sense that any of this makes sense. Viewed from outer space the world is a only insignificant dumpling held together by stale bread. The land of the future, which has been formed into a community on the basis of the whole body of scientific knowledge, is equally terrible. People have had their "competition gland" removed and the result is a monotonous world totally lacking in humor. Someone or other has hanged himself on every tree out of boredom in a totally dismally world.

---

[16] Personal communication from Rudolf Koder, Vienna 1969.

[17] This is a slightly emended English version of the account in Pape (1977, 76f).

A visit with a world famous philosopher (Schopenhauer) is not less cheerless; for he presents a picture of a meaningless world in which joy and sorrow are merely mechanical reactions and ethics an equally meaningless tune played on a flute. All in all we are presented with a totally dismal, entirely sceptical dystopian vision of reality that parodies all concepts of reason, order and progress be it material or moral. After Busch even Schopenhauer is a naive optimist. That was a kind of *Dichtung* that Wittgenstein took to have real philosophical significance.

Eduards Traum (Busch 1959) has a controversial place in the literature on Busch. It is highly unusual in German literature for fitting into no established genre. Some commentators consider it to be a vain effort to establish a reputation in the genre *novella* that in fact is a mere grab-bag full of cranky old Busch's pet peeves. However, other commentators like Josef Kraus (1994) take it to be the high point in his career as a writer, a work almost without precedent that marks the summit of Busch's career. His mode of communication, like Kierkegaard's, is indirect—and clowning—if dismal. There is an essentially absurd—or nonsensical—character to the story, which has contradictory aspects as a short epic. It is an enormously concentrated grotesquery designed to shock us out of our smug complacency and in doing so to expand our horizons.

In his trenchant study of Busch as a powerful critic of classical idealist aesthetics Gottfried Willems (1998) has identified a constellation of ideas and values that go a long way to explaining how the concept of *Dichtung* incorporated in a work like *Eduards Traum* could display the real philosophical urge that Wittgenstein attributed to him. Just as much as Nietzsche, Busch was concerned with tearing down the grand edifice that figures like Winckelmann and Schlegel had built up in the Golden Age of German classicism. His bitter satire is directed at demolishing the idea that literature should be lofty and edifying. It turns on a bitterly ironic scepticism with respect to those ideals, above all, the idea of art as a source of redemption. Busch wants to remind us how petty we really are in our efforts to make ourselves important. He wants to cut us down to size and show us how horribly funny we actually are. Nonsense is his vehicle for bringing us to a disquieting, disappointing form of self-knowledge.

Like Wittgenstein's own mature philosophy *Eduards Traum* culminates in insight but that insight carries with it a certain form of *disappointment*, especially for the kind of philosophical mind that seeks satisfaction or edification from philosophy, the moral high ground or social progress. We know from Wittgenstein's critique of none other than Schopenhauer, who exerted a very powerful influence on him, that Wittgenstein considered clarity of expression of the kind that the Vienna Circle strove for in philosophy to be the mark of a superficial mind even if the subject involved is highly obscure and the writer's view justified. The truly profound, creative thinker allows us to see something that cannot be said at all.[18] Among

---

[18] Commenting upon Wittgenstein's view that Schopenhauer was a superficial thinker Maurice Drury writes, "…a shallow thinker may be able to say something clearly but…a deep thinker makes us see that there is something that cannot be said" (Drury 1981, 96). With respect to his nihilistic pessimism Busch stands exactly in this relation to Schopenhauer (see Willems 1998,

Wittgenstein's heroes, Frege in the *Begriffsschrift* and Heinrich Hertz in the Introduction to *Principles of Mechanics* which Wittgenstein invariably recommended when he wanted to tell his students how philosophy should be done, stand out in this respect.[19] This seems to be what he admired in the grotesque, pessimistic, nonsense in *Eduards Traum*.

Paradoxically, this view of what Wittgenstein most likely admired in Busch can help us in our efforts to understand just what it was that he could appreciate about Kraus and what he may have seen in a work like his monumental play, at once realistic and absurdist, *Die letzten Tage der Menschheit*, which displays many of the characteristics of *Eduards Traum* but on a mega-monumental scale without parallel in world literature. Kraus discovers ultra-revealing quotations in the newspapers and *invents* grotesquery that show the indescribable horrors of war, things which are happening because we cannot imagine them because the very act of imagining them would make the impossible (see Timms 1986, 273–284). Moreover, it was crucial to Kraus that all this be "funny" and we know that Wittgenstein's admiration for Kraus began to wane at precisely the point where he ceased to find him *funny*. The use of nonsensical examples bring us to a profoundly *disappointing* insight that we would prefer to avoid. This is what Wittgenstein *seems* to consider genuine philosophical *Dichten* to be all about. It is something that is an appropriate, if unusual, function of literature but hardly at home in modern analytical philosophy.

That brings us to our final point: Wittgenstein does not and cannot see himself as what we *normally* understand as a poet or even an artist here. The task to which his style has been formed is not the creation of beauty for its own sake but effectively to remove/rob us of self-imposed illusions, inter alia by way of simple but powerful fictions. It was a way of *crafting* philosophy that produces an extraordinary beauty, as Brian McGuinness emphasized in *Young Ludwig* (1988, 77):

> [Philosophy for Wittgenstein] was a craft, a discipline…and its value consisted in its being well done. So one should do it well and not preach about it: …showing not saying was important. Like all crafts, its exercise at its highest produces beauty, a beauty which requires an intellectual effort to grasp …

It was crucial to him to know when to stop in philosophy, just as it was important not to go too far in repairing a defective toilet, as Norman Malcolm learned in the course of Wittgenstein's visit with him at Cornell (Malcolm 1958, 85). This is what separates him from "artist philosophers" like Nietzsche, the later Heidegger or Jacques Derrida: the poetic element in Wittgenstein's work was always subordinated to his philosophical goals without which *dichterische* dimension it would have been radically lacking (as Marjorie Perloff has rightly emphasized). But, for all that, its aim is first, last and always to demonstrate to philosophers that the urge to philosophize is at once perverse and inescapable. Little wonder that disabusing

---

chapter 6 "Macht und Ohnmacht des Geistes: Wilhelm Busch, Schopenhauer, Nietzsche", 80–90).

[19] Hertz's influence upon Wittgenstein, which is second only to Frege's is discussed in detail in Janik (2009).

them of their transcendental illusions had to involve a curious kind of fictionalizing—or as he wrote in the *Big Typescript* (Wittgenstein 2015):

> All the philosophy can do is to destroy idols. And that means not making any new ones—say out of "the absence of idols". (Ts 213, 413r_f)

It is a notion that certainly would have pleased a *Dichter* with a "real philosophical urge" like Wilhelm Busch without being any the less strictly philosophical for all that.

Epilogue

The remarks above were originally inspired by a discussion with Nuno Venturinha at a symposium in Innsbruck commemorating the 60th anniversary of Wittgenstein's death. In that occasion Dr. Venturinha set out to re-contextualize the remark that "Philosophie dürfte man eigentlich nur dichten." In a recent paper on the same topic[20] he makes the following points about the passage:

1. It was originally prefaced by the remark, "the presentation [*Darstellung*] of philosophy can only be poetical."
2. It should be read in conjunction with the following text from 1938: "When I do not want to teach a more correct thought, but another /a new/ movement of thought, then my aim is a 'transvaluation of values' and I come to Nietzsche as to my view that a philosopher should be a poet." [Dr. Venturinha's translation]
3. This implies a Nietzschean inspiration for the idea that *philosophieren* ought to be *dichten*.

My response is to re-iterate that the use of *dichten* does not imply poetry or an opposition to prose but a different way of using prose, fictional prose as opposed to representational prose. The point is one of typically Wittgensteinian subtlety. The fact remains that there are not even hints of any attempts to write poetry in the sense of *Lyrik* anywhere in Wittgenstein's corpus; he just wants to create more powerful ways of showing alternatives to the conventional philosophical imagination. The idea is less Nietzschean in inspiration than it is *parallel* to what Nietzsche wanted to do in ethics. Nothing that Wittgenstein does—always his own ultimate criteria—is very like what Nietzsche does when he is being a poet, say in *Zarathustra*. His continual self-criticism is not the sign of helplessness that it is usually taken to be but a striving for a kind of perfection in the writing of philosophical fiction that would set analytical philosophers on the right track—not one that would put them out of business. If this is right the genuine work of analytical philosophy is fictional but not lyrical and it is every bit as inspired by Frege as it is by the Sage of Sils Maria.

---

[20] "Agrammaticality"; in this volume.

References

Busch, W. (1959). *Eduards Traum*. Stuttgart: Reclam.
Drury, M. O'C. (1981). Some notes on conversations with Wittgenstein. In R. Rhees (Ed.), *Ludwig Wittgenstein: Personal recollections* (pp. 90–111). Totowa: Rowman and Littlefield.
Ficker, L. (2014). *Ludwig (von) Ficker—Ludwig Wittgenstein: Briefwechsel 1914–1920*. A. Steinsiek, & A. Unterkircher (Eds.). Innsbruck: Innsbruck University Press.
Frege, G. (1879). *Begriffsschrift, eine der arithmetischen nachgebildete Formelsprache des reinen Denkens*. Halle: Verlag Louis Nebert.
Frege, G. (1884/1987). *Die Grundlagen der Arithmetik: Eine logisch mathematische Untersuchung über den Begriff der Zahl*. Stuttgart: Reclam.
Frege, G. (1893/1962). *Grundgesetze der Arithmetik, begriffsschriftlich abgeleitet. Bd. 1*. Darmstadt: Wissenschaftliche Buchgesellschaft.
Frege, G. (1989). Gottlob Frege: Briefe an Ludwig Wittgenstein. A. Janik, & C. P. Berger (Eds.). *Grazer Philosophische Studien, 33/34*, 3–33.
Gabriel, G. (1991). Der Logiker als Metaphoriker: Freges philosophische Rhetorik. In G. Gabriel (Ed.), *Zwischen Logik und Literatur: Erkenntnisformen von Dichtung, Philosophie und Wissenschaft* (pp. 65–88). Stuttgart: Metzler.
Jäger, G. (1970). Das Gattungsproblem in der Ästhetik und Poetik von 1780 bis 1850. In J. Hermand & M. Windfuhr (Eds.), *Zur Literatur der Restaurationsepoche 1815–1848* (pp. 2–30). Stuttgart: J.B. Metzlersche Verlagsbuchhandlung.
Janik, A. (2009). *Assembling reminders*. Stockholm: Santérus.
Kant, I. (1966). *Kritik der reinen Vernunft*. Stuttgart: Reclam.
Kraus, J. (1994). *Wilhelm Busch*. Reinbeck bei Hamburg: Rowohlt.
Malcolm, N. (1958). *Ludwig Wittgenstein: A memoir*. London: Oxford University Press.
McGuinness, B. (1988). *Wittgenstein: A life. Young Ludwig, 1889–1921*. London: Duckworth.
McGuinness, B. (2006). In praise of nonsense. In R. Calcaterra (Ed.), *Le ragioni del conoscere e dell'agire: Scritti in onore di Rosaria Egidi* (pp. 357–365). Milan: Franco Angeli.
Pape, W. (1977). *Wilhelm Busch*. Stuttgart: J. B. Metzlersche Verlagsbuchhandlung.
Perloff, M. (2011). Writing philosophy as poetry: Literary form in Wittgenstein. In O. Kuusela & M. McGinn (Eds.), *The Oxford handbook of Wittgenstein* (pp. 714–728). Oxford: Oxford University Press. The article is also to be found on the homepage of the Wittgenstein Initiative, Vienna: http://wittgenstein-initiative.com/
Rothhaupt, J. (2013). Ludwig Wittgenstein über Wilhelm Busch: "He has the REAL philosophical urge.". In V. Munz, K. Puhl, & J. Wang (Eds.), *Language and world: Part two—signs, minds and actions* (pp. 297–315). Heusenstamm: Ontos Verlag.
Sluga, H. (1980). *Gottlob Frege*. London: Routledge and Kegan Paul.
Stern, J. P. (1966). Karl Kraus's vision of language. *The Modern Language Review, 61*, 73–74.
Timms, E. (1986). *Karl Kraus, apocalyptic satirist*. New Haven/London: Yale University Press.
Willems, G. (1998). *Abschied vom Wahren-Schönen-Guten: Wilhelm Busch und die Anfänge der ästhetischen Moderne*. Heidelberg: Universitätsverlag C. Winter.
Wittgenstein, L. (1958). *Philosophical investigations/Philosophische Untersuchungen*. G. E. M. Anscombe (Trans.). Oxford: Basil Blackwell.
Wittgenstein, L. (1967). *Zettel/Zettel*. G. E. M. Anscombe (Trans.). Oxford: Basil Blackwell.
Wittgenstein, L. (1969). *On certainty/Über Gewissheit*. D. Paul, & G. E. M. Anscombe (Trans.). Oxford: Basil Blackwell.
Wittgenstein, L. (1980). *Culture and value/Vermischte Bemerkungen*. G. H. Wright (Ed.), von in collaboration with Nyman, H., P. Winch (Trans.). Oxford: Basil Blackwell.
Wittgenstein, L. (1998). *Culture and value/Vermischte Bemerkungen*. G. H. von Wright (Ed.), in collaboration with Nyman, H.; Rev. ed. by Pichler, A.; P. Winch (Trans.). Oxford: Basil Blackwell.
Wittgenstein, L. (2015–). *Bergen Nachlass Edition* (Wittgenstein Archives at the University of Bergen under the direction of A. Pichler (Ed.)). In *Wittgenstein Source* http://www.wittgensteinsource.org/. Bergen: WAB.

Chapter 10
Agrammaticality

Nuno Venturinha

Abstract In this paper I begin by scrutinizing classic approaches to the question of agrammaticality, with a particular focus on Frege and the early Wittgenstein, and try to show that a further step is needed in order to adequately address this topic. I then focus on the later Wittgenstein's treatment of nonsense-poems and claim that the failure of the *Philosophical Investigations* as a book is actually connected with Wittgenstein's recognition that philosophy should be written under the form of poetry. The therapeutic consequences of this view are discussed in connection with Deleuze's comments on Melville and with Read's comments on Sass, which offer very different views of agrammaticality.

Keywords Agrammaticality • Language • Literature • Nonsense • Philosophy • Wittgenstein

10.1

Philosophy of language is first and foremost concerned with grammar as logical syntax. The agrammatical, i.e. the illogical, is the result of violating the rules of that syntax. There seems to be no room for the agrammatical other than showing instances of what we cannot actually do. However, if such violations take place, this means that they are in a certain way sanctioned by thought. This is paradoxical. I can disobey a traffic sign and take the risk. In doing so, I consciously subject myself to the consequences. But it is hard to conceive of not obeying the rules of thought because they do not offer us, at least at first sight, an alternative. The only consequence of not thinking logically is nonsense. I can hear someone exclaiming: "If you cross a street with the red light for pedestrians, that's nonsense too!" Well, it depends. I may have taken all the precautions by looking both ways and seeing that no vehicle was coming. In this case I will merely be violating the law, not logic. But how can the latter be violated? The same person might reply perhaps more impatiently now: "Have you never observed a misapplication of thought?" Again, it

N. Venturinha (✉)
Universidade Nova de Lisboa, Lisbon, Portugal
e-mail: nventurinha.ifl@fcsh.unl.pt

depends on what we mean by "misapplication of thought". If I say that 2 + 2 = 5 or that the capital of England is Paris, I will merely be committing a mistake in arithmetic or geography. In order to infringe the rules of grammar, I must do something of a different kind—within language.

According to the Wittgenstein of the *Tractatus*, someone who says that "Socrates is identical" violates grammar or logical syntax.[1] This happens because, as he explains in 5.473, "there is no property which is called 'identical'", that is, "because we have not made some arbitrary determination, not because the symbol is in itself unpermissible". In 5.4733 Wittgenstein notes that "we have given *no* meaning [keine *Bedeutung*] to the word 'identical' as *adjective*". A similar example is to be found in Dummett. He observes that "the sentence 'Chairman Mao is rare', while perfectly grammatical, is meaningless because 'rare', though in appearance just like a first-level predicative adjective, has the sense of a second-level predicate" (Dummett 1981, 51). The problem is that the function "rare" requires a first-level function as argument, e.g. "Chairman Mao is revolutionary and that is rare". If we do not follow this rule, Dummett (1981, 50–51) alerts, we will be violating what Frege called the "hierarchy of types" and Russell the "theory of types".

Both "Socrates is identical" and "Chairman Mao is rare" differ substantially from Heidegger's repeated employment of the concept "nothing" and in particular of the expression "The nothing itself nihilates" in his inaugural public lecture in Freiburg in 1929.[2] In a well-known article from 1931, Carnap strongly criticizes Heidegger for mistakenly using "nothing" [*Nicht*] as a noun and "nihilates" (other translations are "nothings" or "noths") [*nichtet*] as a verb.[3] Carnap (1959, 69) takes Heidegger's expressions as illustrative of "metaphysical pseudo-statements of a kind where the violation of logical syntax is especially obvious, though they accord with historical-grammatical syntax". As to "nothing", Carnap (1959, 70) says that "it is customary in ordinary language to use it in this form in order to construct a negative existential statement", like "Nothing is outside". However, it is "the fabrication of the meaningless word 'to nothing'" that Carnap (1959, 71) finds more outrageous. In his opinion, "the meaningless words of metaphysics usually owe their origin to the fact that a meaningful word is deprived of its meaning through its metaphorical use in metaphysics" and "here we confront one of those rare cases where a new word is introduced which never had a meaning to begin with" (Carnap 1959, 71). As Carnap remarks a bit later on, Heidegger was "clearly aware of the conflict between his questions and statements, and logic". It is not without relevance that Heidegger always refers to logic within quotation marks. He knows that its laws are not being observed and this is exactly his aim. Contrary to Carnap, Heidegger thinks that there is still a "domination of 'logic' in metaphysics", with the term

---

[1] Cf. Wittgenstein (1922, 5.473 and 5.4733). See also 3.325, where Wittgenstein speaks of "rules of *logical* grammar—of logical syntax", and 3.344, where he simply refers to "rules of logical syntax". The notion of "logical syntax" also appears in 3.33 and 6.124.

[2] Cf. Heidegger (1998, 84ff).

[3] Cf. Carnap (1959, 69–73).

"logic" corresponding, as he makes clear in the first edition of the text, to "traditional logic and its *logos* as origin of the categories" (Heidegger 1998, 95). The upshot is obvious: "The idea of 'logic' [as "the *traditional* interpretation of thinking"] itself disintegrates in the turbulence of a more originary questioning" (Heidegger 1998, 92). This is unacceptable for Carnap who maintains that Heidegger is under the illusion that he is thinking *something* with the "pseudo-statements" he uses.

Here we are close to the author of the *Tractatus* for whom Frege, more than anyone else, had seen the fundamental issue but not in the right light. Wittgenstein thus opposes Frege's view according to which a "legitimately constructed proposition must have a sense [*Sinn*]" to the idea that a "possible proposition is legitimately constructed, and if it has no sense this can only be because we have given no *meaning* [*keine* Bedeutung] to some of its constituent parts" (Wittgenstein 1922, 5.4733). He points out parenthetically: "Even if we believe that we have done so."

Incidentally, that was not exactly what Frege said. In his "On Sense and Meaning" he writes: "It may perhaps be granted that every grammatically well-formed expression figuring as a proper name always has a sense [*Sinn*]. But this is not to say that to the sense there also corresponds a thing meant [*Bedeutung*]" (Frege 1892/1980, 28/58). And even if here, as Kevin Klement notes, "Frege limits his discussion of the *Sinn/Bedeutung* distinction to 'complete expressions'", what happens is that these include "names purporting to pick out some object *as well as whole propositions*" (Klement 2002, 11, my emphasis). In his analysis "concerning the sense and reference for an entire declarative sentence [*Behauptungssatz*]", Frege (1892/1980, 32/62) avers that whereas the "sense" corresponds to the "thought" [*Gedanke*] behind the sentence, the "reference" stands for its "truth value" [*Wahrheitswert*]. Thus he asks: "Is it possible that a sentence as a whole has only a sense, but no reference [*keine Bedeutung*]?" (Frege 1892/1980, 32/62). Frege's example of our imagining Odysseus doing so and so illustrates that an individual "representation" [*Vorstellung*] can be communicated to others under the form of a "thought", if the proposition has "sense", but that a "reference" does not necessarily accompany the "sense". This is where the "truth value" enters. If, as Frege (1892/1980, 33/63) recognizes, it is unimportant to determine the *Bedeutungen* of Homer's *Odyssey* "so long as we accept the poem as a work of art", the scenario radically changes if we become interested in the "truth" of what is at issue. Frege thus seems to be aware that we can construe propositions in general which apparently have "sense" but "no reference"—with this resulting in Wittgensteinian nonsensicality, in regard to which philosophers must be vigilant.

It is worth mentioning that the antepenultimate proposition of the *Tractatus* states that the only admissible method in philosophy would be to "say nothing except what can be said, *i.e.* the propositions of natural science", that is to say, "something that has nothing to do with philosophy", and "when someone else wished to say something metaphysical, to demonstrate to him that he had given no meaning [*keine Bedeutung*] to certain signs in his propositions" (Wittgenstein 1922, 6.53). This is the reason why Wittgenstein astonishingly claims in 6.54 that his own (metaphysical) propositions are "nonsensical" [*unsinnig*]. They are in the service of elucidation and hence need to be overcome. Instead of our trying to find a determinate *thesis* throughout the book, be that positivist, metaphysical or resolute,[4] the *Tractatus* is probably best read as resulting in a genuine perplexity for which silence becomes the most reasonable attitude. The so-called "Dictation for Schlick" from the early 1930s is illuminating in this regard. In commenting specifically on Heidegger's "What is Metaphysics?" Wittgenstein said:

> If we want to deal with a proposition such as "The nothing noths" [...] we ask ourselves: What did the author have in mind with this proposition? [...] Our method resembles psychoanalysis in a certain sense. To use its way of putting things, we could say that a simile at work in the unconscious is made harmless by being articulated. [...] But how is it possible to demonstrate to someone that this simile is actually the correct one? This cannot be shown at all. But if we free him from his confusion then we have accomplished what we wanted to do for him. (Wittgenstein and Waismann 2003, 69–71)

The way Wittgenstein discusses the use of similes and therapy cannot be interpreted as if they were required solely for the understanding of someone else's statements. Our very way of conceiving reality is indeed determined by a constellation of similes, the majority of which lie unquestioned as to their nature. Carnap is in fact misled by the idea that only metaphysicians are responsible for metaphysical claims. As Kant rightly noticed, a *metaphysica naturalis* is already at work under the form of "a natural disposition" (1998, B21). That is why Kant wants to establish its possibility as a science. Wittgenstein also attempts to define its limits but he puts the emphasis on language. Waismann's notes from a conversation that took place at the end of 1929 show a Wittgenstein confessing that he "can imagine what Heidegger means by being and anxiety", that one "feels the urge to run up against the limits of language" or, citing Kierkegaard, "against paradox", and that this "is *ethics*". He goes on:

> I think it is definitely important to put an end to all the claptrap about ethics [. . .]. In ethics we are always making the attempt to say something that cannot be said, something that does not and never will touch the essence of the matter. [...] But the inclination, the running up against something, *indicates something*. St. Augustine knew that already when he said: What, you swine, you want not to talk nonsense! Go ahead and talk nonsense, it does not matter! (Wittgenstein 1979a, 68–69)

In the "Dictation for Schlick", Wittgenstein introduces an interesting notion, that of an "inarticulate sound". In his view, there is the temptation to justify what is puzzling for us through even more puzzling images. What he says is that "we are accustomed to calm our mental disquiets by tracing certain propositions back to more fundamental ones", with these consisting of "an inarticulate sound" in the end (Wittgenstein and Waismann 2003, 75).[5] Wittgenstein's conclusion is that "a proposition such as 'The nothing noths' is in a certain sense a substitute for this sort of

---

[4] For a discussion of the two traditional readings of the *Tractatus*, the positivist and the metaphysical, in contrast with the resolute reading of the book, see, for example, Conant (2002). In this paper, Conant, after Diamond (1991), discusses at length some of the passages I have alluded to.

[5] See in addition Wittgenstein (2009, §261), where he also refers to the idea of "inarticulate sound".

inarticulate sound", corresponding in the domain of an author's activity to "a requirement of style" (Wittgenstein and Waismann 2003, 75). Wittgenstein does not thus seem to be against metaphysical talk like Heidegger's, if one is fully aware of its nonsensicality—as Heidegger was. What can be seen from that point of view is not at stake, since nothing can meaningfully be seen from there. What really matters is the impulse, "the inclination", to see things from an entirely new point of view and not just from a different one.

Let us now consider another species of nonsense, Lewis Carroll's famous "Jabberwocky", also alluded to by Wittgenstein. Having quoted part of it in his very first lecture in Cambridge in 1930, Wittgenstein (1980, 3) said that this poem "can be analysed into subject and predicate and parts of speech". Carroll does not create neologisms to which an innovative meaning can be attached and henceforth discussed. He deliberately produces meaningless words combining them with others that we are acquainted with and observing the basic rules of grammar. The result is nonsensical only because *we have given no meaning to some of its constituent parts*. If we were taught what each of Carroll's signs mean, we could make sense of the whole poem provided that its grammatical structure is maintained. Wittgenstein summarizes this well in the *Big Typescript*, contemporary with the "Dictation for Schlick", when he writes: "What is a proposition?—First of all, there is a *sound* to the propositions in our language. (Hence nonsense poems like those of Lewis Carroll.) And often what *we* call nonsense is not something that's arbitrary" (Wittgenstein 2005: 59, translation slightly modified).[6] Some years later, rewriting this passage, Wittgenstein would round off the discussion observing that "when *we* say in different places that this and that have no sense, what does have no sense is never an arbitrary combination of words and syllables, but something that *seems* to have a sense, that *sounds* like a proposition or part of a proposition" (Wittgenstein 2000, MS 116, 59, my translation).[7] In short, all the cases analysed so far still obey some principles of grammar. No single example was of a truly agrammatical portion of text. But can we go further than that? Let us look more closely at what poetry can do, again following in Wittgenstein's footsteps.

10.2

In the *Philosophical Investigations*, after having introduced the builders' language, Wittgenstein mentions Carroll. He writes:

If we say, "Every word of language signifies something", we have so far said nothing *whatever*; unless we explain exactly *what* distinction we wish to make. (It might be, of course, that we wanted to distinguish the words of language (8) from words "without meaning"

---

[6] See in addition Wittgenstein (2009, §134), where it is said apropos of the sentence "This is how things are" that "*one* feature of our concept of a proposition is *sounding* like one" (translation slightly modified).

[7] References follow G. H. von Wright's catalogue of the Wittgenstein papers and are by MS or TS number.

such as occur in Lewis Carroll's poems, or words like "Tra-la-la" in a song.) (Wittgenstein 2009, §13, translation slightly modified)

The sentence "Every word of language signifies something", which is a translation for "*jedes Wort der Sprache bezeichnet etwas*", is puzzling.[8] According to Peter Hacker in his revised edition of the commentary on the *Investigations* written with Gordon Baker, what Wittgenstein had in mind was "the language of §8" referred to later in the remark (Baker and Hacker 2005, 66). Following this line of reasoning, Hacker and Schulte have opted for "in the language" in their revised translation of the text, adding the definite article to Elizabeth Anscombe's translation. My reason for adopting "of language" is that Wittgenstein himself has crossed out the definite article in Rush Rhees' translation of the very same phrase as "every word of the language denotes something" in the so-called "early" version of the *Investigations* (Wittgenstein 2000, TS 226, 8).[9] I am thus of the opinion that what Wittgenstein meant was language in general. Additional evidence is given by Wittgenstein's handwritten insertion of "our" before "language" in the Rhees typescript when he mentions the section number. This is important because what Wittgenstein seems to be criticizing is the (philosophical) idea that every word must possess some meaning. If we do not know the meaning of a word, we simply do not know the uses it can have. But what about "words without meaning"? It is worth noting that Rhees' original translation was "'nonsense' words" and that Wittgenstein was more literal rendering "*ohne Bedeutung*" by "without meaning". To be sure, words cannot be nonsensical; what can be nonsensical are combinations of words into sentences—be they elliptical or not.[10] There is a remark in which Wittgenstein speaks of "nonsensical words" [*Unsinnsworte*] when saying that "a poet's words can pierce me through and through", but he certainly envisaged the whole poem because he also observes that we could say that "these words sound full of meaning" (Wittgenstein 2000, MS 180b, 2v–3r, my translation). What is relevant for our purposes is that a nonsensical combination of words does produce an effect on us. A "square circle", for example, points to the impossibility of conceiving *that*. So, when Wittgenstein talks about "words without meaning", like Carroll's, he is certainly not eschewing them.

In what is probably the best comparison ever made between Wittgenstein and Carroll, George Pitcher refrains from commenting on §13 and puts forward the arguable view that their common concern was with "the kind of nonsense that results from the very natural confusions and errors that *children* might fall into, if only they were not so sensible" (1986, 414). I think Pitcher undervalues Carroll but it is Wittgenstein who interests me now. Pitcher goes as far as to argue that "Wittgenstein, being a philosopher, exerted all his efforts to drag us back to reality

---

[8] This is actually reminiscent of what Frege says in §31 of the *Basic Laws of Arithmetic*, namely: "Our simple names denote [*bedeuten*] something" or "always have a denotation [*Bedeutung*]" (1893/1964, 47/87). See in this regard Weiner (2002).

[9] This typescript consists of a partial translation of the first half of the "early" or "pre-war" version of the *Investigations*, thoroughly revised by Wittgenstein. The German text is to be found in TS 220, 8, deriving from MS 142, 10.

[10] See Wittgenstein (2009, §§19–20).

from the (horrible) world of myth and fantasy" (Pitcher 1986, 414). There must be something wrong here. If the nonsense at stake is one proper to "*children*", why should philosophy care about it? Are we all like Alice, living a fantastic adventure? Maybe, but it is hard to believe Wittgenstein wanted to cure "us" of such a confusion. Contrary to what Pitcher thought, Wittgenstein was much in favour of nonsense. In addition to the above-quoted remarks, I would like to quote this dictum from a few years before Wittgenstein's death: "Don't *for heaven's sake*, be afraid of talking nonsense! Only don't fail to pay attention to your nonsense" (Wittgenstein 1998, 64). So far as I can see, the last thing Wittgenstein is interested in is an expurgation of nonsense for the simple reason that it belongs to, and manifests, human nature. His aim is, as he sums up in the *Investigations*, "to pass from unobvious nonsense to obvious nonsense" (Wittgenstein 2009, §464). In effect, if meaning is use, as Wittgenstein maintains, the employment of meaningless words creates some result. This is completely different from believing that a meaning can be assigned to every word. Thus Andrew Lugg writes in his commentary:

> In [§13] Wittgenstein is trying to get us to look askance at the apparently anodyne philosophical assumption that language can be—and needs to be—explained by referring to what its words "*signify*". […] The important thing to notice is that Wittgenstein is attempting to disperse some of the fog surrounding our thinking about the meaning of words. (Lugg 2004, 35–36)

Wittgenstein's discussion of fairy tales in the *Investigations* is in this regard illuminating. He says:

> "But in a fairy tale a pot too can see and hear!" (Certainly; but it *can* also talk.)
> "But a fairy tale only invents what is not the case: it does not talk *nonsense*."—It's not as simple as that. Is it false or nonsensical to say that a pot talks? Does one have a clear idea of the circumstances in which we'd say of a pot that it talked? (Even a nonsense-poem is not nonsense in the same way as the babble of a child.) (Wittgenstein 2009, §282a–b, translation slightly modified)[11]

The contrast Wittgenstein establishes here between the nonsense produced by a certain poem and that arising from a child babbling makes it plain that there are for him varieties of nonsensicality, with nonsense acting sometimes as *sense*. In one of his notebooks, he writes: "Think again about how much sense there is in a nonsense-poem!" (Wittgenstein 2000, MS 127, 51, my translation). This "sense" is not the result of a permissible combination of meaningful words. In a nonsense-poem, neither are all words meaningful nor do they always combine in a coherent way. The "sense" is given by the experience we have when confronting the poem—a "mystical" experience in the early Wittgenstein's terminology. In a remark penned in the early 1930s that would make its way into the *Big Typescript* we find: "How about sentences that occur in poems [*Dichtungen*]? Here one certainly can't speak of verification, and yet these sentences have sense." (Wittgenstein 2005, 69,

---

[11] In an early version of this remark, there is a variant for the beginning of the second paragraph, which reads: "But the fairy tale only speaks *falsehood*; not *nonsense*!" (2000, MS 124, 239, my translation)

translation slightly modified)[12] A distinctive feature of poetry is its capacity to suggest views from nowhere. These depend on the reader's imagination to freely play, as Kant would have it, with the content of the poem. The deeper the poet *de-constructs* language, the more his or her readers participate in this game. They are forced to abandon their presuppositions about the common meanings of words and to instinctively reflect upon new relations of concepts, be they possible or not. Colours, smells, sounds, textures, everything enters into the poem making each word signify in a flash. It does not matter whether *it* can be true; *it is*, we feel, and during these moments we are transported to a different world in which words have *their meaning*. I agree with Donald Davidson that there is no hidden meaning to be grasped when one moves back and forward through the metaphors of a poem. What they mean is "what the words, in their most literal interpretation, mean, and nothing more" (Davidson 1984, 245). If poetry were simply a more sophisticated way of conveying the same cognitive content that standard uses of language can convey, then it would lose all its significance. As Davidson (1984, 261) emphasizes, it is erroneous to believe "that a metaphor carries a message, that it has a content or meaning (except, of course, its literal meaning)". He recognizes that a metaphor suggests an "insight" but in his view most of it "is not propositional in character" (Davidson 1984, 263).

This experience is described by Wittgenstein in a later version of the remark that contained the expression "nonsensical words". This he dropped, rewriting the whole text as follows:

> A poet's words can pierce us through and through. And that is of course *causally* connected with the use that they have in our life. And it is also connected with the way in which, conformably to this use, we let our thoughts roam up and down in the familiar surroundings of the words. (Wittgenstein 1981, §155, translation slightly modified)

What a poet does is to play with ordinary language suggesting visions. He or she is not interested in naming or in the logical validity of his or her sentences. The poet's commitment is to art. As pointed out by Wittgenstein in a 1940 note, "a poet, when he poetizes [*dichtet*], makes a psychological experiment"; he goes on: "Only *so* can it be explained that a poem can have a value." (Wittgenstein 2000, MS 117, 196, my translation)[13] This has consequences. The first relates to what we mean by understanding something. We often think that the sign of having understood this or that is to be able to give a satisfactory account of it. But there is no satisfactory account to be given in regard to a poem. Any attempt to interpret its content leads to incom-

---

[12] Luckhardt and Aue render "*Dichtungen*" by "works of literature", but "poems" seems to capture more adequately what Wittgenstein has in mind here.

[13] This remark occurs in the context of a critical discussion of "calculation as experiment" and within quotation marks, suggesting that Wittgenstein would separate poetry from experiment. Smyth (1997, 228) alludes exactly to "Wittgenstein's distinction between poetry and 'psychological experiment'", but he also writes that "Wittgenstein's analogous distinction between prediction and proof, experiment and calculation, did not prevent him from considering proofs predictions or *metapredictions* about future events". Indeed, Wittgenstein seems to accept that the poet makes a kind of psychological experiment when he writes a poem.

pleteness in a twofold way: (i) incompleteness of the multiple relations an interpreter can project when he or she is affected; (ii) incompleteness of the expressions he or she can use in order to describe his or her affectations. I shall call the first *incompleteness of interpretation* and the second *incompleteness of expression*. They are the two great enemies of literary criticism. The difficulty further increases when one faces nonsense-poems. Already in the course of revising the *Big Typescript*, Wittgenstein had remarked: "Do we *understand* Christian Morgenstern's poems, or Lewis Carroll's poem 'Jabberwocky'? In these cases it's very clear that the concept of understanding is a fluid one." (Wittgenstein 1974, 43).[14]

If we pay attention to other remarks of Wittgenstein's on understanding, we may feel tempted to conclude that although poetry holds a creative power, it is not the best candidate for philosophical inquiry. In fact, it is not—if we take philosophical inquiry to mean conceptual clarification. True, this is the sort of inquiry philosophers can do. From the time of the *Tractatus,* however, Wittgenstein moves between philosophy and literature and my view is that his later thought is far more ambiguous than the earlier. The argumentative strategies Wittgenstein uses are not only meant to make us see what philosophy can reveal but also what lies outside its scope. The *Philosophical Investigations*, appearances to the contrary notwithstanding, is one among other collections of remarks in the *Nachlass* aimed at showing that there is no definite line of reasoning to be followed, no ultimate theory (or therapy) to be learned. Philosophical analysis can shed light on who we are, how we lead our lives, but it stops at some point. That is exactly where philosophy meets poetry. In no other place in Wittgenstein's writings is this tension better described than in the following observation from 1933 or 1934:

> I believe I summed up where I stand in relation to philosophy when I said: really one should write philosophy only as one *writes a poem* [*Philosophie dürfte man eigentlich nur* dichten]. That, it seems to me, must reveal how far my thinking belongs to the present, the future, or the past. For I was acknowledging myself, with these words, to be someone who cannot quite do what he would like to be able to do. (Wittgenstein 1998, 28)

With only a few notable exceptions,[15] commentators have not paid attention to the Nietzschean inspiration of this confession. In 1938, when Wittgenstein was about to submit the so-called "early" version of his new book for publication, we find this remark which he crossed out:

> When I do not want to teach a *more correct* thought, but another /a new/ movement of thought, then my aim is a "transvaluation of values" and I come to Nietzsche as to my view that the philosopher should be a poet [*Dichter*]. (Wittgenstein 2000, MS 120, 145r, my translation)

As late as 1947, Wittgenstein would continue to worry about the weakness of his philosophical "prose" and say that he felt as incapable doing what he wanted with it

---

[14] In a manuscript version of this remark, Wittgenstein wrote "Do we finally *understand* [. . .]" (2000, MS 140, 6, my translation).

[15] See Schalkwyk (2004, 72–73), Schroeder (2006, 125–126), Brusotti (2009, 346) and Schulte (2013, 349–352).

as he felt incapable of writing "a poem". He ends by recognizing that "[i]t is like someone's saying: In this game I can attain only *this* level of perfection, and not *that*" (Wittgenstein 1998, 67).

This certainly sounds strange to those who see Wittgenstein as the champion of rule-following, as if his philosophical mission consisted simply of eliminating conceptual confusion. To show that a large amount of philosophical debate involves nonsensical talk was surely one of Wittgenstein's lifelong concerns. However, when his posthumous writings are carefully taken into account, what we find is an author struggling with language in all its dimensions, including the performativity of philosophical discourse. Little wonder that the remark in which Wittgenstein originally said that "one should write philosophy only as one *writes a poem*" is prefaced by a more general note saying: "The presentation of philosophy can only be poetized [*Die Darstellung der Philosophie kann nur gedichtet werden*]" (Wittgenstein 2000, MS 115, 30, my translation). These claims cannot be oversimplified for what Wittgenstein states is not that *he* wished to be a poet but that poetry is the adequate "form of presentation" for philosophical matters. Of course, in this way philosophy, more than losing scientificity, loses space as a discipline. Wittgenstein's failure to arrange the *Philosophical Investigations* as a book where, as written in the preface, "the thoughts should proceed from one subject to another in a natural, smooth sequence [*in einer natürlichen und lückenlosen Folge*]" (Wittgenstein 2009, 3) is the immediate consequence of such a view.[16] It is highly significant that so many scholars write on Wittgenstein but no one has adopted his heterodox literary style, still less the poetic discourse he suggests. Doing Wittgensteinian philosophy should actually mean, as Nietzsche and the *Tractatus* had already recommended, *overcoming philosophy*.

10.3

The outcome of this examination may seem disappointing. A philosophical inquiry into agrammaticality results in the affirmation of literature, in particular of poetry, as the *medium* through which we can apprehend what is the most important thing for us: a real discovery of ourselves. As a matter of fact, we do not make effective progress when we attempt to theorize about what has always been the case. We gain knowledge about what we do, both theoretically and practically, and develop theses in accordance with that diagnosis, but twenty-five centuries have passed and philosophy has still not answered the questions it was supposed to answer. Maybe it cannot answer those questions. But what then is its relevance? From Wittgenstein's point of view, philosophy is a propaedeutic activity. As a scientific activity, it is bound to fail. It should be taught in order to lead people to see beyond the customary way of looking at the world but at the same time to surpass that standpoint. Teaching

---

[16] Wittgenstein's words remind us of those of Frege in his *Foundations of Arithmetic* when it is said that "only if every gap [*Lücke*] in the chain of deductions is eliminated with the greatest care can we say with certainty upon what primitive truths the proof depends" (1884/1980, §4).

that this is how things are while keeping the same perspective, with philosophy providing continuous illumination, is admittedly trifling. If we want to take a step forward, working not with *more correct thoughts* but with *other* or *new movements of thought*, we must cease to think of grammar and logic as impregnable.

This programme is not different in nature from psychoanalysis. The difference lies in that a text takes the place of a patient. Wittgenstein, who was deeply influenced by Freud, alludes in the early version of the *Investigations* to "a *theory* (a 'dynamical' theory of the proposition, etc.)", which is not a real "theory" but "a form of presentation" [*Form der Darstellung*] capable of grasping "something which lies beneath the surface" (Wittgenstein 2000, TS 226, 71).[17] Although this view can be seen as a criticism of the Tractarian picture theory and of our unwarranted processes of generalization,[18] Wittgenstein actually endorses this dynamism. He tried to practise it until the end of his life, doing justice to what he had stated in the *Big Typescript*: "Thought is dynamic" (Wittgenstein 2005, 126). However, even "sometimes jumping, in a sudden change, from one area to another", his "philosophical remarks" could never quite accomplish their goal: "to travel criss-cross in every direction over a wide field of thought" (Wittgenstein 2009, 3). The *Investigations* offer mere approximations towards this goal. Literature does it better because only it can solve what Deleuze in the preface to his *Essays Critical and Clinical* calls "the problem of writing". He explains that "writers, as Proust says, invent a new language within language, a foreign language, as it were". He goes on to say:

> They bring to light new grammatical or syntactic powers. They force language outside its customary furrows, they make it *delirious*. [...] in effect, when another language is created within language, it is language in its entirety that tends toward an "asyntactic", "agrammatical" limit, or that communicates with its own outside. [...] Literature is a health. (Deleuze 1998, lv)

The writer is a therapist, not in the way the New Wittgensteinians employ the term but as someone offering an understanding of the movement of life. We are usually so absorbed in our mundane activities that the situation we are in is rarely taken into consideration. Sometimes life forces us to do this but the everyday course of events returns. We can protest for better working conditions or in favour of democracy, but our existence in the world seems to be naturally justified. Writers take *this* protest in hand. As in many other protests, we cannot expect an immediate solution. It is the identification of the phenomena that matters.

---

[17] Talking about dreaming, Wittgenstein ascribes the idea of a "dynamical theory" to Freud in MS 157a, 56v, in a note from the beginning of 1937, and explicitly mentions Freud's "'dynamic' theory of dreams" in TS 239, 74. This remark in the so-called "revised early version" of the *Investigations* would make its way into the collection of cuttings published as *Zettel* (§444). Compare this with the aforementioned passage from Wittgenstein's "Dictation for Schlick", in which he says that "[o]ur method resembles psychoanalysis".

[18] See, for example, Stern (1995, 39–40).

Melville's *Bartleby*, as Deleuze notes, is one of these cases where literature takes the lead in protesting on behalf of humanity. In this short story, the author reveals much more than the state of mind of a peculiar character called Bartleby. What Melville reveals is the apparent meaninglessness of human life as *we* usually lead it. Bartleby acts as a mirror for any of us, representing what Wittgenstein, in the same place where he talks of "'dynamical' theory", calls a "prototype of *all* cases" (Wittgenstein 2000, TS 226, 71). Again, this is a generalization but it points to something. Deleuze's analysis focuses on a phrase repeatedly employed by Bartleby: "I would prefer not to."[19] As Deleuze observes, the more current usage is "I had rather not" and he concludes:

> Certainly it is grammatically correct, syntactically correct, but its abrupt termination, NOT TO, which leaves what it rejects undetermined, confers upon it the character of a radical, a kind of limit-function. [...] Murmured in a soft, flat, and patient voice, it attains to the irremissible, by forming an inarticulate block, a single breath. In all these respects, it has the same force, the same role as an *agrammatical* formula. (Deleuze 1998, 68)[20]

According to Deleuze (1998, 76), what this dictum produces is "a zone of indetermination or indiscernibility" caused by the consciousness of the value of life. Bartleby is not tired of a certain task, he does not reject this or that; his answers are full rejections. This existential mood is well described by the Wall Street lawyer at the beginning of the story when he says: "What my own astonished eyes saw of Bartleby, *that* is all I know of him, except, indeed, one vague report which will appear in the sequel" (Melville 1996, 21–22). The use of italics in this passage is another stylistic resource used by Melville to provoke a vibration in language.[21] Instead of being uttered by the author, we are left with the void that arises from this image. What Bartleby finds out is that his job, as scrivener, is ironical in the face of life. He becomes conscious of the inevitability of death and regards our ordinary practices, which we do our best to preserve, as worthless. Bartleby's nothingness cannot be seen as the result of vocational failure or some sort of depression. He is surely depressed but he is not a particular case to be studied from a clinical point of view; he is, so to speak, "the prototype of *all* cases", as is, for example, Hofmannsthal's Lord Chandos or Pessoa's heteronymous Bernardo Soares. It is the illness of life that makes its appearance and no answer seems to be enough. Deleuze comments on this illness as follows:

> Illness is not a process but a stopping of the process, as in "the Nietzsche case". Moreover, the writer as such is not a patient but rather a physician, the physician of himself and of the world. [...] Literature then appears as an enterprise of health. (Deleuze 1998, 3–4)

---

[19] Cf. Melville (1996, 32ff).

[20] Compare the notion of "inarticulate block" with Wittgenstein's notion of "inarticulate sound" alluded to above.

[21] Compare Wittgenstein's similar procedure in the above-quoted remark about "perfection" (1998, 67).

10.4

In a paper titled "On approaching schizophrenia through Wittgenstein", Rupert Read offers a quite different view of nonsense. Read contends against Louis Sass' use of Wittgenstein's notion of solipsism as a key for interpreting the schizophrenic mind in his *The Paradoxes of Delusion*. Reflecting on the famous case of Daniel Paul Schreber, author of *Memoirs of My Nervous Illness*, Sass writes:

> [. . .] as Wittgenstein knew very well, demonstrating the logical impossibility and incoherence of solipsism as a philosophical doctrine can hardly preclude its existence in the actual world, both as an explicitly held belief (philosophical solipsism proper) and as an implicitly felt mood (what I am calling quasi-solipsism). [. . .] And, he seemed to think, although this intuition could not really be *said* (because it was nonsensical, tautologous), it could in some sense be *shown*—by pointing to the mood, attitude or form of life in which the doctrine is rooted. (Sass 1994, 75)

Read calls attention to "Schreber's frustration with the inadequacies of language to express what he wants to say", pointing to "striking coinages of Schreber's own, such as the neologism, 'fleeting-improvised-man'" (Read 2001, 452). Read's view is that "*there can be no such thing as understanding schizophrenia*" (Read 2001, 466–467). He claims:

> What is it like to be schizophrenic? It's quite literally not (literally) *like* anything. (But yet Sass still wants to tell us what it is like …) (Read 2001, 467)

And Read concludes:

> You cannot use Wittgenstein to gain *insights into* nonsense—and problem cases of schizophrenia *centrally involve* the problematics of dealing with nonsense. (Read 2001, 467)

I agree with Read that the Wittgensteinian solipsist cannot be seen that way, but for different reasons. Read's slogan is "Solipsism is nonsense" and nonsense for him, following Cora Diamond's and James Conant's teaching, "is nothing" (Read 2001, 449 ff.). He is totally against the idea that, for example, Antonin Artaud's writings "can be understood—can be successfully *interpreted*—via Wittgenstein's philosophy" (Read 2001, 451).[22] I concede that to identify a discourse like Artaud's as solipsistic in the sense that it is separated from the world, constituting a "private language", is erroneous, but it is another thing to say that it is "nothing". Such identification is erroneous because solipsism is a term used by Wittgenstein to indicate the limit a subject experiences when confronted with the fact that each of us is the actual condition of existence of the world. This is not to argue that the subject creates his or her world; it is instead to recognize that external objects and other minds only exist within *my* projection. That is why Wittgenstein says in the *Tractatus* that "solipsism strictly carried out coincides with pure realism" (Wittgenstein 1922, 5.64). There is not an inner without an outer as there is not an outer without an inner. However, the realization that this is so is uncommon. We usually look at our position in the world as if we were another thing "among others, among beasts, plants, stones, etc., etc.", as Wittgenstein (1979b, 82) observes on the very same day when

---

[22] It is noteworthy that Artaud is the author of "An Antigrammatical Effort Against Lewis Carroll", which consists of a peculiar translation of "Jabberwocky" into French.

he originally made the remark about solipsism coinciding with realism. To recognize myself as a "philosophical I", as a "metaphysical subject", and not simply as a "human being", is what Wittgenstein means by solipsism.[23] And if we follow his remarks in their original sequence, it becomes clear that this understanding is far from being epistemologically or theoretically orientated.[24] The background in which Wittgenstein moves is *ethical* in such a way that his talk about solipsism assumes the same agrammatical role that we find exemplified in Melville's *Bartleby*. The dissolution of pseudo-problems that is characteristic of Wittgenstein's later philosophy can only be fully apprehended within such a perspective.

But let us come back to Read. He is surely right when he stresses that we should not try to make sense of schizophrenic talk as of another culture "in terms other than its own" (Read 2001, 457). Yet is there *really* an alternative? Read believes that Peter Winch's considerations in his well-known paper "Understanding a primitive society", in which he defends description over interpretation, furnishes a good model. But any description already involves an interpretation. Of course we shall never know what someone like Artaud was going through when he said and wrote the things we know have been said and written by him. Anthropology too is full of examples where imposition of our rules on alien societies has turned out to be a mistake. But can I know with certainty, for instance, what a friend or a relative of mine is experiencing if I am told—or if I notice—that he or she has fallen in love? On the other hand, what am I supposed to do if I observe a member of a tribe offering flowers to another and at the same time blushing? Would it be mistaken to explain this case by appealing to our pattern of blushing with shame when we show our feelings?[25] In order for me to describe behaviours, an interpretation must already be in place. As Wittgenstein remarks over and over again, behaviourism is an illusion. Read's view of nonsense undercuts the possibility of understanding anything that cannot be verified by means of experience—and that is a lot! I am not claiming that schizophrenic language in particular involves the same indeterminacy as ordinary language. What I am saying is that the objectivity of ordinary language, in the midst of which schizophrenic language-games can be recognized and interpreted as constituting a different form of life, depends on subjectivity.[26]

It is this same subjectivity that enables us to grasp an agrammatical passage though we do not find the means to explain it objectively to someone else. Read's argumentation gives a vivid sign of weakness when he asserts that Wittgenstein denies there are "different kinds of nonsense" but at the same time directs the reader to the following endnote:

---

[23] See Wittgenstein (1979b, 92), as well as Wittgenstein (1922, 5.641).

[24] A study of this kind can be found in Venturinha (2011).

[25] I find Read's affirmation that "Winch looks at what the Azande are doing *with and alongside* their words [. . .]" (2001, 458) incomprehensible.

[26] For Read (2001, 472, n. 37), they do not constitute an alternative "way of being-in-the-world"; they form "various ways of ... *not* being-in-the-world *at all*". But his ellipsis shows the opposite.

> This is not contradicted by (e.g.) PI 282. My point is not to make the dogmatic—nonsensical!—*assertion* that there really is only one kind of nonsense, but to suggest that no category of "profound nonsense" or "nonsense in virtue of an incompatibility between the component-words the utterance" [*sic*] will, to use Diamond's words, ultimately be found satisfactory by one. (Read 2001, 473, n. 50)

I think it is only reasonable to admit that when Wittgenstein says in §282 of the *Investigations* that "a nonsense-poem is not nonsense in the same way as the babble of a child" or when he draws attention to the "sense" which "a nonsense-poem" contains, he is in fact contradicting a reading such as Read's. *There are indeed varieties of nonsense* and some of them seem to be the only "form of presentation" of what we count as "profound" in our lives. One can follow Read's suggestion and reserve the word "understanding" for "contexts in which there is a reasonably clear distinction between understanding and *not* understanding someone" (Read 2001, 469) assuming that *I* correctly understood someone else or that *I* have been understood by someone else. As the later work of Wittgenstein demonstrates, the criteria we use rest on very complex presuppositions such as, for instance, that I am not being cheated. But one can also do as Humpty Dumpty suggested to Alice and put both of our eyes on the left or on the right side of our nose or then our mouth above them: "Wait till you've tried" (Carroll 1994, 96).[27]

References

Baker, G. P., & Hacker, P. M. S. (2005). *Wittgenstein: Understanding and meaning: Volume 1 of an analytical commentary on the Philosophical Investigations, exegesis §§1–184*. P. M. S. Hacker (Ed.). Oxford: Blackwell.
Brusotti, M. (2009). Wittgensteins Nietzsche: Mit vergleichenden Betrachtungen zur Nietzsche-Rezeption im Wiener Kreis. *Nietzsche-Studien, 38*, 335–362.
Carnap, R. (1959). The elimination of metaphysics through logical analysis of language. A. Pap, (Trans.). In A. J. Ayer (Ed.), *Logical positivism* (pp. 60–81). New York: Free Press.
Carroll, L. (1994). *Through the looking-glass*. London: Puffin Books.
Conant, J. (2002). The method of the *Tractatus*. In E. H. Reck (Ed.), *From Frege to Wittgenstein: Perspectives on early analytic philosophy* (pp. 374–462). Oxford: Oxford University Press.
Davidson, D. (1984). What metaphors mean. In D. Davidson (Ed.), *Inquiries into truth and interpretation* (pp. 245–264). Oxford: Clarendon Press.
Deleuze, G. (1998). *Essays critical and clinical*. D. W. Smith & M. A. Greco (Trans.). London: Verso.
Diamond, C. (1991). Frege and nonsense. In C. Diamond (Ed.), *The realistic spirit: Wittgenstein, philosophy, and the mind* (pp. 73–79). Cambridge, MA: MIT Press.
Dummett, M. (1981). *Frege: Philosophy of language*. London: Duckworth.

---

[27] Material included in this paper was presented in events in Innsbruck and Lisbon in 2011, Strasbourg in 2012, Athens in 2013 as well as in Lisbon again in 2015. I would like to thank the participants in these meetings for stimulating discussions. I would also like to thank Andrew Lugg, Gisela Bengtsson, Simo Säätelä and Rob Vinten for helpful comments. This paper was written as a contribution to the research project "Wittgenstein's *Philosophical Investigations*: Re-Evaluating a Project" (2010–2013) funded by the Portuguese Foundation for Science and Technology.

Frege, G. (1884/1980). *Die Grundlagen der Arithmetik: Eine logisch mathematische Untersuchung über den Begriff der Zahl*. Breslau: W. Koebner. English edition: Frege, G. (1980). *The foundations of arithmetic: A logico-mathematical enquiry into the concept of number* (2nd rev. ed., 5th impr. with corr.). J. L. Austin (Trans.). Oxford: Basil Blackwell.

Frege, G. (1892/1980). Über Sinn und Bedeutung. *Zeitschrift für Philosophie und philosophische Kritik, 100*, 25–50. English edition: Frege, G. (1980). On sense and meaning. M. Black, (Trans.). In P. Geach & M. Black (Eds.), *Translations from the philosophical writings of Gottlob Frege* (pp. 56–78). Oxford: Blackwell.

Frege, G. (1893/1964). *Grundgesetze der Arithmetik: Begriffsschriftlich abgeleitet* (Vol. I). Jena: Hermann Pohle. English edition: Frege, G. (1964). *The basic laws of arithmetic: Exposition of the system*. M. Furth (Ed. & Trans.). Berkeley: University of California Press.

Heidegger, M. (1998). What is metaphysics? D. F. Krell, (Trans.). In M. Heidegger, *Pathmarks* (pp. 82–96). W. McNeill (Ed.). Cambridge: Cambridge University Press.

Kant, I. (1998). *Critique of pure reason*. P. Guyer & A. W. Wood (Ed. & Trans.). Cambridge: Cambridge University Press.

Klement, K. C. (2002). *Frege and the logic of sense and reference*. New York: Routledge.

Lugg, A. (2004). *Wittgenstein's Investigations 1–133: A guide and interpretation*. London: Routledge.

Melville, H. (1996). Bartleby, the scrivener: A story of Wall-Street. In H. Melville (Ed.), *The Piazza tales* (pp. 15–52). New York: The Modern Library.

Pitcher, G. (1986). Wittgenstein, nonsense, and Lewis Carroll. In S. Shanker (Ed.), *Ludwig Wittgenstein: Critical assessments* (Vol. 4, pp. 398–415). London: Routledge.

Read, R. (2001). On approaching schizophrenia through Wittgenstein. *Philosophical Psychology, 14*, 449–475.

Sass, L. (1994). *The paradoxes of delusion: Wittgenstein, Schreber, and the schizophrenic mind*. Ithaca: Cornell University Press.

Schalkwyk, D. (2004). Wittgenstein's "imperfect garden": The ladders and labyrinths of philosophy as *Dichtung*. In J. Gibson & W. Huemer (Eds.), *The literary Wittgenstein* (pp. 55–74). New York: Routledge.

Schroeder, S. (2006). *Wittgenstein: The way out of the fly-bottle*. Cambridge: Polity Press.

Schulte, J. (2013). Wittgenstein on philosophy and poetry. In M. F. Molder, D. Soeiro, & N. Fonseca (Eds.), *Morphology: Questions on method and language* (pp. 347–360). Bern: Peter Lang.

Smyth, J. V. (1997). A glance at SunSet: Numerical fundaments in Frege, Wittgenstein, Shakespeare, Beckett. In B. H. Smith & A. Plotnitsky (Eds.), *Mathematics, science, and postclassical theory* (pp. 212–242). Durham: Duke University Press.

Stern, D. (1995). *Wittgenstein on mind and language*. New York: Oxford University Press.

Venturinha, N. (2011). Wittgenstein reads Nietzsche: The roots of Tractarian solipsism. In E. Ramharter (Ed.), *Unsocial sociabilities: Wittgenstein's sources* (pp. 59–74). Berlin: Parerga.

Weiner, J. (2002). Section 31 revisited: Frege's elucidations. In E. H. Reck (Ed.), *From Frege to Wittgenstein: Perspectives on early analytic philosophy* (pp. 149–182). Oxford: Oxford University Press.

Wittgenstein, L. (1922). *Tractatus logico-philosophicus*. C. K. Ogden (Trans.). London: Routledge and Kegan Paul.

Wittgenstein, L. (1974). *Philosophical grammar*. R. Rhees. (Ed.), A. Kenny (Trans.). Oxford: Blackwell.

Wittgenstein, L. (1979a). *Wittgenstein and the Vienna Circle: Conversations recorded by Friedrich Waismann*. B. McGuinness (Ed.), J. Schulte & B. McGuinness (Trans.). Oxford: Blackwell.

Wittgenstein, L. (1979b). *Notebooks 1914–1916*. G. H. von Wright & G. E. M Anscombe (Eds.), G. E. M. Anscombe (Trans.). Oxford: Blackwell.

Wittgenstein, L. (1980). *Wittgenstein's lectures: Cambridge, 1930–1932*. D. Lee (Ed.). Oxford: Blackwell.

Wittgenstein, L. (1981b). *Zettel*. G. E. M. Anscombe & G. H. von Wright (Eds.), G. E. M. Anscombe (Trans.). Oxford: Blackwell.

Wittgenstein, L. (1998). *Culture and value: A selection from the posthumous remains* (2nd ed.). G. H. von Wright with H. Nyman (Ed.), A. Pichler (Rev. Ed.), P. Winch (Trans.). Oxford: Blackwell.

Wittgenstein, L. (2000). *Wittgenstein's Nachlass: The Bergen electronic edition*. Oxford: Oxford University Press.

Wittgenstein, L. (2005). *The Big Typescript: TS 213*. C. G. Luckhardt & M. A. E. Aue (Eds. & Trans.). Oxford: Blackwell.

Wittgenstein, L. (2009). *Philosophical investigations*. P. M. S. Hacker & J. Schulte (Eds.), G. E. M Anscombe, P. M. S. Hacker, & J. Schulte (Trans.). Oxford: Wiley-Blackwell.

Wittgenstein, L., & Waismann, F. (2003). *The voices of Wittgenstein: The Vienna Circle*. G. Baker (Ed.), G. Baker, M. Mackert, J. Connolly, & V. Politis (Trans.). London: Routledge.

Author Index

A
Allison, H.E., 70
Alnes, J.H., 3, 4, 24, 39, 106, 115
Anderson, R.L., 61
Anscombe, G.E.M., 88, 89
Appelqvist, H., 71, 72
Aristotle, 10, 86, 112, 149
Artaud, A., 171, 172
Aschenbrenner, K., 120
Aue, A.E., 166
Augustine, 162

B
Baker, G.P., 164
Bar-Elli, G., 24, 29
Baumann, J.J., 148, 153
Beardsley, M.C., 20
Benacerraf, P., 24
Bengtssson, G., 30, 31, 44, 72, 89, 98, 173
Betti, A., 63
Black, M., 2, 35, 76
Boltzmann, L., 152
Bolzano, B., 147
Boole, G., 25, 27
Bradley, F., 54
Bronzo, S., 98
Brusotti, M., 167
Burge, T., 5, 54, 63, 64, 68, 92, 93
Busch, W., 153–156

C
Calderón, P., 11
Carl, W., 29, 36–38, 42

Carnap, R., 10, 81, 120, 133, 134, 149, 160–162
Carroll, L., 11, 163, 164, 167, 171, 173
Cerbone, D., 49
Cervantes, M. de, 11
Church, A., 137, 138
Conant, J., 105, 106, 108, 162, 171
Costreie, S., 26

D
Davidson, D., 88, 89, 166
de Jong, W.R., 63
Deleuze, G., 169, 170
Derrida, J., 155
Diamond, C., 91, 106, 108, 162, 171, 173
Dickens, Ch., 128, 138
Dingler, H., 130
Doyle, C., 133, 137, 138
Drury, M.O'C., 154
Dummett, M., 26, 29, 34, 131, 132, 147, 160

E
Eder, G., 37
Erdmann, B., 148
Ernesti, J.C.T., 16
Euclid, 68

F
Ficker, L., 146
Floyd, J., 3, 38, 43, 106, 115

Frege, G., 1–6, 10, 11, 24–42, 44, 47, 76, 87–98, 101–115, 119, 121, 124, 126, 128, 130, 137, 139, 143, 146, 147, 155, 160, 161, 164, 168
Friedman, M., 56, 57, 64–67

G
Gabriel, G., 4, 10–21, 30, 60–62, 102, 103, 108–110, 121, 122, 128, 133, 139, 153
Goethe, J.W. von, 18, 150
Goldfarb, W., 106, 109
Gumbrecht, H.U., 20

H
Haaparanta, L., 52, 115
Hacker, P.M.S., 164
Hänsel, L., 144
Heck, R., 26, 34
Hegel, G.W., 10
Heidegger, M., 102, 150, 155, 160–163
Heine, E., 26, 78, 81
Helmholtz, H. von, 53
Hertz, H., 53, 71, 155
Hilbert, D., 14, 37, 113
Hofmannsthal, H. von, 11, 170
Homer, 13, 109, 123, 133, 138, 161
Husserl, E., 124
Hyder, D., 51, 53, 60

I
Ishiguro, H., 113

J
Jäger, G., 145
Janik, A., 5, 6, 143, 144, 147, 152, 155
Jeshion, R., 24, 29, 33, 68

K
Kant, I., 4, 18, 35, 37, 41, 49–65, 67–71, 106, 113, 147–149, 152, 162, 166
Kanterian, E., 5, 94, 96
Keller, G., 18, 19
Kierkegaard, S., 88–90, 102, 154, 162
Kitcher, P., 24

Klement, K.C., 161
Kluge, E.-H.W., 76
Koder, R., 153
Kraus, J., 154, 155
Kraus, K., 151, 152
Kremer, M., 94
Künne, W., 5, 96, 121, 133, 134, 137, 139

L
Leibniz, G.W., 57, 149
Lewis, D., 138
Linnebo, Ø., 27
Lotze, H., 60, 61
Luckhardt, C.G., 166
Lugg, A., 165, 173
Lukács, G., 18

M
MacFarlane, J., 30, 48
Mahr, B., 41
Malcolm, N., 155
Mayr, K., 145
McGuinness, B., 151, 152, 155
Meinong, A., 13
Melville, H., 170, 172
Methlagl, W., 146
Mezzadri, D., 47, 49
Michaëlis, C., 41
Moore, G.E., 1, 138
Morgenstern, Chr., 167
Mörike, E., 153
Musil, R., 21

N
Neurath, O., 144
Nietzsche, F., 102, 154–156, 167, 168, 170

P
Pape, W., 153
Parsons, T., 13, 121, 133, 134, 139
Patton, L., 71
Peano, G., 15
Perloff, M., 145, 146, 155
Pessoa, F., 170
Pichler, A., 145, 146
Pitcher, G., 164, 165
Plato, 10, 12, 102, 147
Polimenov, T., 5, 103, 119, 134

Author Index

Q
Quine, W.V.O., 29, 102
Quintilianus, M.F., 16

R
Read, R., 171–173
Reck, E.H., 56
Resnik, M.D., 81
Rhees, R., 151, 153, 164
Richards, I.A., 10
Ricketts, T., 29–31, 56, 57, 92, 109, 112
Rilke, R., 146, 150
Rothhaupt, J., 153
Russell, B., 1, 2, 15, 36, 52, 53, 69–71, 79, 81, 82, 87, 111, 128, 149, 160
Ryle, G., 138, 152

S
Säätelä, S., 44, 173
Salmon, N., 36
Sass, L., 171
Savigny, E. von, 149
Scaliger, J.C., 16
Schalkwyk, D., 167
Schildknecht, C., 17, 20
Schiller, F., 110, 127, 128
Schlegel, F., 145, 154
Schlick, M., 162, 163, 169
Schliemann, H., 137, 139
Schopenhauer, A., 71, 152, 154
Schreber, D.P., 171
Schröder, E., 25, 27, 148
Schroeder, S., 145, 167
Schulte, J., 164, 167
Schulze, J.G., 70
Searle, J., 11, 12, 49, 120–122, 125, 128, 130, 132
Shakespeare, W., 128
Shapiro, S., 24, 29, 31, 33
Sidney, P., 10
Sluga, H., 147, 148
Smith, N.J.J., 5, 90, 94, 95
Smyth, J.V., 166
Stenius, E., 71

Stenlund, S., 4, 26, 76
Stern, D., 169
Stern, J.P., 151
Sullivan, P., 69
Sundholm, G., 41
Sutrop, M., 20

T
Tang, H., 69
Taschek, W.W., 5, 48, 49, 93, 94
Textor, M., 121
Thomae, J., 4, 26, 76–81, 83–85
Timms, E., 155
Tolstoy, L., 19, 20

V
van Heijenoort, J., 103, 106
Venturinha, N., 6, 156, 159–173
Vieta, F., 80, 81, 86
Vinten, R., 173
Voltaire, F., 128

W
Waismann, F., 1, 78, 79, 162
Walsh, W.H., 50
Wang, J., 145
Webb, J., 70
Weiner, J., 3, 24, 29, 30, 34, 42, 104, 106, 109, 112, 115, 164
Willems, G., 154
Winch, P., 144, 145, 172
Windelband, W., 60–62
Wittgenstein, L., 1–6, 17, 21, 53, 54, 65, 67–72, 77–83, 87–91, 94, 97, 98, 102, 103, 105, 108, 113, 137, 143–145, 148, 154, 155, 160–173
Wittgenstein, P., 152
Wolff, Ch., 16
Wright, G.H. von, 1, 2, 6, 102, 145, 163

Z
Zola, E., 10

Subject Index

A

Accommodating understanding of the reader, 17
Agrammaticality, 6, 156, 159, 168
Algebra, algebraic, 25, 27, 28, 80
Analytic/synthetic distinction, 37
A priori/a posteriori distinction, 4, 14, 24, 36, 37, 40, 41
Arithmetic, arithmetical, 3–5, 15, 25–28, 30, 31, 34, 35, 41, 43, 44, 52, 63, 66, 67, 76–79, 81–86, 103, 105, 147, 148, 152, 160, 164, 168
As-if assertion, *see* Pseudo-assertion
Assembling reminders, 144, 150
Assertion (*Behauptung*), 5, 11–13, 15, 19, 20, 24, 28, 30, 32, 37, 38, 42, 56, 57, 59, 62, 88, 90–92, 94–97, 104, 109–111, 115, 120, 122, 123, 126, 128, 131, 133, 137–139, 149, 173
Assertion-sign, 87, 88, 94, 95, 131
Assertoric discourse, 11–13
Assertoric force (illocutionary force, *behauptende Kraft*), 3, 11, 12, 15, 48, 89–91, 95–97, 102, 109, 120, 122, 130, 131
Assertoric sentence, 110, 122, 123, 125, 129–132
Autonomy, 61
Axiom, 4, 15, 25, 28, 31–35, 37–40, 42–44, 49, 76, 114
Axiomatic system, 26, 28, 31, 32, 38, 43, 44, 147

B

Begriffsschrift, 4, 5, 14, 15, 24–28, 31, 32, 34, 35, 39–43, 63–67, 71, 87, 89–91, 94–98, 103, 105, 106, 108, 109, 112, 115, 139, 146, 147, 155

C

Calculus, 25, 65, 66, 79, 82
Categorial, 16, 17
Categories, 50–52, 64, 67, 69, 120, 124, 149, 161
Causality, 33
Certainty, 50, 152, 168, 172
Cognitio ex principiis, 63
Cognitive value, 10, 13, 16–21, 35, 37, 128
Cognitivism, cognitivistic, 4, 10, 13, 17
Colouring of thought (*Färbung*), 12, 15, 122
Completeness, 28, 53, 66, 79
Concept-formation, 56, 77, 150
Concept-script, *see Begriffsschrift*
Conditio humana, 21
Conditional-stroke, 27, 28, 30, 32, 39
Consciousness, 54, 55, 57, 58, 61, 70, 71, 170
Content, 5, 12–14, 16, 17, 19, 21, 24–28, 30, 32–37, 39–41, 43, 50, 51, 56–59, 64, 70, 76, 77, 79, 83–85, 89–96, 102, 113, 114, 122, 124, 126, 129, 130, 136, 137, 139, 153, 166
Content-stroke, 27, 28, 30, 91
Continental tradition, 144
Convention, 29, 40, 89, 95, 98

Craftsmanship, 143
Customary reference (*gewöhnliche Bedeutung*, extensional contexts), 13, 120
Customary sense (*gewöhnlicher Sinn*), 13, 133

D

Definition, 2, 5, 15, 30, 33, 34, 37–39, 43, 79, 107, 109, 112, 113, 122, 123, 128
Description, 1, 3, 12, 13, 17, 19, 20, 29, 39, 43, 63, 65, 69, 70, 81, 90, 122, 123, 128, 153, 172
Dichtung (*dichterische Sprache, dichterischer Gebrauch, Dichtkunst, Dichter*, poetry, poem, poet, poetic, eloquence, fiction, fictionality, fictive, fictional discourse, fictional text, narrative, literary, artistic, *Lyrik, Drama, Prosa*), 2, 11, 87–90, 101–103, 121, 145, 165
Disappointment, 154
Dream, 11, 30, 153, 169
Dynamical theory, 169, 170

E

Elucidation (*Erläuterung*), 3–5, 16, 17, 24, 30, 32, 33, 39, 40, 43, 101, 103–105, 107–109, 112–115, 162
Emotivism, emotivistic, 4, 10, 11, 13, 17, 21
Empiricism, 26, 35, 57
Epistemic value, 35–38, 42, 43
Epistemology, 21, 24, 25, 31, 40
Equinumerosity, 79
Example, 4, 5, 10, 13–16, 18, 19, 26, 27, 30, 33, 35, 37, 38, 42, 51, 58, 67–70, 85, 89–91, 93, 96, 102, 107–112, 120, 121, 123, 124, 126, 127, 129, 130, 132, 135, 136, 138, 139, 144, 147, 148, 150–152, 155, 160, 162–164, 170–172
Experience, 19, 49–51, 54, 55, 57, 58, 63, 128, 149, 165, 166, 171, 172
Expressibility, 147, 167
Extensional contexts, *see* Customary reference

F

Fact, factual reality, 3, 5, 6, 10–12, 14–16, 19, 21, 26, 27, 33, 35, 36, 43, 48, 51, 56–59, 61–63, 65–68, 70, 76, 82, 85, 88, 89, 93–95, 97, 108, 113, 125, 127–129, 134, 135, 137–139, 144–148, 150–152, 154, 156, 160, 162, 167, 168, 171, 173

Fairy-tales, 13, 165
Feeling (phantasy, *Vorstellung*), 10, 16, 20, 21, 53, 56–58, 110, 114, 122, 124, 161, 172
Fictional sentence, 123, 127
Fictive object, fictive entity, 12–14
Figurative expression, 17, 108, 114
Formalism, 4, 5, 26, 29, 31, 76, 78–80, 83, 84
Function, 4, 10, 11, 13, 15–18, 21, 39–41, 50–52, 56, 60, 63–67, 69, 70, 76, 81–84, 88–90, 92, 93, 97, 107, 108, 111, 112, 124, 147, 148, 155, 160, 170
Functor, 93

G

Geometry, geometrical, 25–28, 30–32, 35, 37, 42, 43, 64, 68–71, 79, 103, 123, 124
Gesture, *see* Hint
Grammar, 71, 104, 107, 145, 148, 151, 159, 160, 163, 169

H

Hint (*Wink*, gesture), 2, 3, 16, 17, 114, 115, 147, 156
Humanities, 2, 3, 6, 15, 102, 115, 170
Humor, 152, 153

I

Idea (*Vorstellung*), 13, 16, 18, 21, 24, 27, 29, 37, 49, 53, 56, 58, 60–62, 65, 67, 70, 76, 77, 79, 83, 85, 88, 90, 91, 93–95, 104, 121, 133–135, 146–150, 152–154, 156, 161, 162, 164, 165, 169, 171
Identity, 30, 33–39, 69, 109
Illness, 170, 171
Illocutionary force, *see* Assertoric force
Imagination, 20, 21, 70, 145, 149–151, 156, 166
Inarticulate sound, 162, 163, 170
Incompleteness of expression, 167
Incompleteness of interpretation, 167
Indirect reference (*ungerade Bedeutung*, intensional contexts), 13, 36, 120, 126, 133, 134
Ineffability, 70
Inference, 14, 15, 26–28, 31, 32, 40, 43, 52, 63, 67–69, 96, 103
Influence, 14, 24, 29, 32, 37, 60, 61, 81, 103, 121, 143, 146, 147, 152, 154, 155, 169

Subject Index

Intensional contexts, *see* Indirect reference
Intensional object, 13, 14, 134, 136, 137
Intuition (*Vorstellung*), 40, 41, 49–51, 58, 64, 69–71, 147, 152, 153, 161, 171

J

Judgment
    analytic, 4, 30, 38, 40–42, 44
    synthetic, 4, 14, 15, 24, 35–37, 39–41
Judgment-stroke, 5, 27, 28, 30, 42, 43, 87–98

K

Kinematics, 53
Knowledge, 1, 10, 15, 19–21, 35–38, 41, 42, 52, 54, 62–67, 76, 77, 101, 106, 128, 147, 150, 153, 168

L

Language game, 132, 138, 148, 149, 151
Literature, 3, 5, 10–13, 16, 18, 20, 21, 24, 47, 102, 110, 120, 133, 134, 144, 145, 152–155, 166–170
Logic
    Boolean, 25, 26
    constitutivity of, 48
    as doctrine, 65, 68, 71
    formal, 63, 76, 147
    justification of, 52
    laws of, 1, 4, 30, 48, 49, 56, 61, 62, 65, 66
    normativity of, 47–72
    pure general, 51, 61
    as a science, 24–44, 104
    as a theory, 63, 65
    universalist conception of, 67, 94
Logic$_F$, 65–67, 69
Logic$_T$, 65–67
Logical inference, 15, 37, 103
Logical norms, 61
Logical objects, 24, 52
Logical structure, 24, 53, 94
Logical syntax, 81, 159, 160
Logicism, 15, 24, 25, 27, 29, 33, 34, 43, 44, 61

Logic of discovery, 15
Logic of justification, 15

M

Mathematics, 2, 3, 25–27, 30, 35, 37, 38, 40, 63, 64, 68, 76, 78, 80, 81, 83, 84, 86, 95, 104, 111, 112, 119, 148
Meaning (*Bedeutung*), 2, 4, 6, 10, 17–19, 21, 24, 28–30, 32, 35–37, 39, 40, 55, 58, 62, 72, 78, 81, 83, 84, 102, 104, 106, 113, 114, 122, 137, 143, 145, 160, 161, 163–166. *See also* Reference
Metalanguage, 17
Metamathematics, 81
Metaphysics, 4, 29, 30, 56, 69, 76, 144, 160
Misinterpretation, 19
Mock-thought, *see* Pseudo-thought

N

Name, 5, 12, 13, 18, 29, 34, 35, 42, 60, 77, 80, 83, 91–94, 97, 106, 107, 110–112, 121–128, 130, 133, 134, 137–139, 161, 164
Necessity, 33, 51, 53, 55, 60, 68, 95, 107, 114
Negation, 27, 28, 30, 33, 35, 41, 57, 95, 108, 127, 132
Negation-sign, 27, 28
Neo-Kantian, neo-Kantianism, 52–54, 57, 60–62
Non-psychological conceptualism, 14
Nonsense collection, 151
Nonsense, nonsensical, nonsensical words (*Unsinn, unsinnig, Unsinnsworte*), 3, 6, 68, 69, 105, 107, 108, 151, 152, 154, 155, 159, 162–168, 171–173
Nonsense-poem, 6, 163, 165, 167, 173
Numbers, 11, 26, 29–31, 34, 35, 40, 52, 55, 67, 68, 76–79, 81, 82, 119, 121, 150, 152, 153, 163, 164

O

Objectivity, 29, 54, 56, 59, 61, 62, 65, 105, 147, 172
Object language, 17
Ontological object, 14

P

Perception, 20, 42, 49, 54, 55, 57, 77, 81, 149
Phantasy, *see* Feeling
Philosophical instruments, 104, 133, 149
Philosophical method, 6, 143, 144, 147, 148
Philosophical practice, 145, 150, 151
Philosophical temptation, 149, 152
Philosophical wit, 153
Philosophy, 1, 2, 6, 10–12, 14, 17, 24, 31–42, 49, 53, 57, 64, 65, 71, 72, 76, 83, 84, 91, 101–103, 106, 108, 112–115, 119, 120, 143–156, 161, 165, 167–169, 172
Pictorial form, 70, 71
Platonism, 56, 61
Predicate, 5, 13–15, 50, 52, 58, 91–93, 97, 107, 108, 123–126, 133–136, 139, 160, 163
Private language argument, 149
Propaedeutic, 2, 3, 107, 168
Proposition, 1, 3, 14, 19, 26, 29, 32, 34, 37, 38, 41–44, 53, 68–72, 83, 86, 88, 93, 94, 97, 103, 107, 108, 111, 113, 143, 147, 153, 161–163, 169
Propositional content, *see* Thought
Protreptic, 17
Pseudo-assertion (as-if assertion, *Scheinbehauptung*), 111, 120, 127–132
Pseudo-proper name (*Scheineigenname*), 126, 127
Pseudo-thought (mock thought, *Scheingedanke*), 12, 126–129
Psychoanalysis, 162, 169
Psychologism, 29, 48, 54, 61, 109

R

Rationality (*Vernunft*), 59, 63–67, 70, 101, 105, 106
Reality, 11, 19, 21, 61, 69, 128, 150, 154, 162, 164
Real object, 125, 128, 133, 135
Reference (*Bedeutung*), 2, 11–14, 16, 17, 19, 21, 37, 48, 49, 54, 55, 60, 76, 78, 79, 81–84, 89, 93, 102, 104, 109–114, 120, 122–130, 133–138, 145–147, 160, 161, 164. *See also* Meaning
Reference-less singular term, 126
Referential meaning, 18

Referring, 21, 128, 165
Re-presentation (*Vergegenwärtigung*), 11, 15, 17, 19–21, 32, 50, 51, 54, 55, 57–65, 69–71, 96, 102–104, 108, 110–112, 127, 148, 149, 161
Rules
  constitutive, 48–52
  normative, 48, 60

S

Saying, 17, 21, 51, 57, 66, 76, 80, 83, 91, 92, 97, 106, 107, 114, 136, 150, 155, 164, 168, 172
Schizophrenia, schizophrenic, 171, 172
Science
  exact, 15, 102, 115
  logic as, 24
  natural, 151, 161
  and philosophy, 10, 12, 31–42, 102
  (*see also Wissenschaft*)
Seemingly true (*scheinbar wahr*), 121, 125
Self-evidence, 33, 38, 39, 69
Self-knowledge, 154
Sense (*Sinn*), 5, 12, 25, 48, 76, 88, 106, 120, 145, 160
Sensibility, 50–52, 54, 60, 63, 64, 68, 70
Shading (*Beleuchtung*), 15, 16
Showing, 10, 11, 21, 155, 156, 159, 167
Sign, 5, 11, 25, 26, 28, 30, 34, 38, 39, 41–43, 68, 71, 72, 76, 78–85, 87–92, 94–98, 110, 111, 114, 123, 124, 130–132, 135, 150, 151, 156, 159, 161, 163, 166, 172
Singular terms, 24, 123–125, 127, 133–136, 138
Solipsism, 171, 172
Speech act, 5, 11, 12, 20, 90, 121–123, 129, 130, 132
Spontaneity, 60
Subjectivism, subjectivist, 54
Subjectivity, subjective, 16, 29, 50, 53–56, 58–60, 113, 172
Symbolic meaning, 18, 19
Symbol, symbolic, 5, 18, 19, 21, 26, 27, 32, 35, 78, 80–83, 86, 95, 147, 160

T

Theory of types, 81, 82, 160
Thesis of fictional discourse, 20
Thought
 illogical, 48, 69, 71, 72
 laws of, 33, 34, 48, 53, 105, 106
Thought (*Gedanke*, propositional content), 3, 12, 24, 47, 76, 87, 102, 120, 144, 159
Transcendental deduction, 50, 51, 55, 64, 65
Transcendental philosophy, 64, 65
Truth, 1, 10, 26, 48, 86, 94, 101, 121, 143, 161
Truth-claim, 122, 129, 132
Truth-value, 5, 14, 33, 35, 36, 38, 42, 43, 54, 76, 92, 97, 111, 122, 124–126, 129, 134–136
Truth-valueless sentence, 122, 126, 128, 130

U

Understanding, 2, 3, 6, 12, 14, 16, 17, 20, 24, 25, 27, 30, 35–39, 41, 42, 48–52, 55, 57–62, 64, 68, 79, 82, 101, 102, 104, 106, 107, 109, 112–115, 135, 144, 148, 162, 166, 167, 169, 171–173
Unreality, 147

V

Vienna Circle, 78–80, 144, 154

W

Wissenschaft (science, scientific discourse, scientific philosophy), 1, 10, 24, 51, 76, 101, 122, 150, 161

CPSIA information can be obtained
at www.ICGtesting.com
Printed in the USA
LVHW081812090619
620633LV00006B/101/P